QUANTUM FIELD THEORY AND BEYOND

Essays in Honor of
Wolfhart Zimmermann

QUANTUM FIELD THEORY AND BEYOND

Essays in Honor of Wolfhart Zimmermann

Proceedings of the Symposium in Honor of Wolfhart Zimmermann's 80th Birthday

Ringberg Castle, Tegernsee, Germany 3 – 6 February 2008

editors

Erhard Seiler
Max-Planck-Institut für Physik, Germany

Klaus Sibold
Universität Leipzig, Germany

World Scientific

NEW JERSEY • LONDON • SINGAPORE • BEIJING • SHANGHAI • HONG KONG • TAIPEI • CHENNAI

Published by

World Scientific Publishing Co. Pte. Ltd.
5 Toh Tuck Link, Singapore 596224
USA office: 27 Warren Street, Suite 401-402, Hackensack, NJ 07601
UK office: 57 Shelton Street, Covent Garden, London WC2H 9HE

British Library Cataloguing-in-Publication Data
A catalogue record for this book is available from the British Library.

QUANTUM FIELD THEORY AND BEYOND
Essays in Honor of Wolfhart Zimmermann
Proceedings of the Symposium in Honor of Wolfhart Zimmermann's 80th Birthday

ISBN-13 978-981-283-354-9
ISBN-10 981-283-354-4

Printed in Singapore by Mainland Press Pte Ltd.

PREFACE

This volume collects twelve essays written in honor of Wolfhart Zimmermann on the occasion of his 80th birthday. Ten of them are based on talks given at a Symposium in his honor, held at the Ringberg castle of the Max Planck Society from February 3rd to 6th, 2008.

Wolfhart Zimmermann has been in the forefront of research in Quantum Field Theory since the 1950s, when the famous work of 'LSZ' (Lehmann, Symanzik, Zimmermann) was created, which is at the basis of all modern applications of Quantum Field Theory to Accelerator Physics, and without which results expected from the LHC at CERN – scheduled to start operation this year – would not be possible. But he was also the first person to construct composite operators in Quantum Field Theory, thus laying the groundwork for the mathematical description of symmetries and their breaking in perturbation theory with its undisputable success in the standard model of particle physics. On the more abstract level this led to the understanding of anomalies: all anomalies known up to date have their origin in an identity which he proved for normal products of different subtraction degrees. In particular anomaly coefficients are given in terms of the coefficients of this identity. The range of this observation has not yet been fully exhausted until the present day, but remains a source of new structural relations.

The perturbatively formulated operator product expansion which he established gave safe ground to the corresponding non-perturbative conjecture of Kenneth Wilson which lead to quantitatively successful results in Quantum Chromodynamics.

His study of asymptotic freedom in the case of several coupling constants introduced the concept of reduction of couplings which enlarges the notion of symmetry and under mild assumptions builds a bridge from perturbation theory to the non-perturbative regime of a theory. The proof that theories exist which have vanishing dilatation and conformal anomalies to all orders of perturbation theory is based on this concept. Amongst these models is the famous supersymmetric Yang-Mills theory with four supersymmetries.

Some of the contributors of this volume were his students, some were his collaborators, some just shared his interest in the subject, but all of them are inspired by him and his work. Analogously one might classify the contributions by content. Some continue work originally initiated by him and employ his tools directly. Some are devoted to lay new groundwork in quantum physics or to broaden and to deepen applications. But all are close in spirit and originality. Two of the contributions are particularly remarkable by the fact that they employ ideas and concepts of Quantum Field Theory in different areas of physics, such as the mathematical theory of electrons in disordered media and the theory of dynamical systems. They thus prove how fruitfully renormalization theory has transcended its origin in particle physics and leads to new insight in otherwise seemingly disjoint parts of physics.

A glance at the subjects covered also reveals the enormous richness and diversity of a theory that originally aimed at a mathematically sound theory of elementary particles. New concepts of spacetime are being checked, fundamentals of quantum mechanics are formulated; Quantum Field Theory is embedded in new structures. The essays in this volume attest both to Wolfhart Zimmermann's inspiring influence and the power and continuing vigor of Quantum Field Theory in our days.

We have tried to arrange the contributions roughly according to increasing distance from Wolfhart Zimmermanns own work in Quantum Field Theory; of course the linear order required by the presentation cannot do justice at all to the various interconnections between the different articles.

We would like to thank all contributors for their carefully written essays which provide an entertaining tour through part of todays theoretical physics.

For funding the symposium thanks are due to the Max Planck Institute for Physics in Munich and to the Max Planck Institute for Mathematics in the Sciences in Leipzig. We are most grateful to A. Hörmann and his team for perfect organization and warm hospitality at the Ringberg castle. Finally we would like to thank T. Hahn for providing the picture on which the cover is based and P. Breitenlohner for technical help with the preparation of this volume.

Erhard Seiler — Munich, Germany
Klaus Sibold — Leipzig, Germany

31 July 2008

CONTENTS

ZIMMERMANN'S SUBTRACTION SCHEME AND THE PERTURBATIVE SOLUTION TO THE RENORMALIZATION GROUP EVOLUTION EQUATIONS

CARLO BECCHI*

Università di Genova, Dipartimento di Fisica
and I.N.F.N. Sezione di Genova
via Dodecaneso 33, Genova
I-16146, Italy
**E-mail: becchi@ge.infn.it*

In the framework of Euclidean field theory we show that an infrared safe slightly modified version of Zimmermann's subtraction scheme generates the perturbative solutions to the Wilson-Polchinski renormalization group equations.

Keywords: Wilson-Polchinski Renormalization Group; BPHZ renormalization

1. Introduction

On the occasion of Wolfhart Zimmermann's 80th birthday I think that a short look at the present status of Quantum Field Theory is certainly timely. I would like in particular to give an example of the persisting fundamental role of many Zimmermann's contributions in the development of Quantum Field Theory.

No doubt quantum field theory is one of the major achievements of twenty's century physics.[1] Even if no interacting four dimensional model has yet been solved, an axiomatic framework leading to a well defined scattering theory is now clearly defined and different constructive approaches have been set up for a class of models. Lehmann-Symanzik-Zimmermann construction of scattering amplitudes has been and remains a basic step in the construction of a complete theory. Among the constructive methods the most important are loop ordered perturbative renormalization[2] and Wilson's renormalization group (R.G.).[3] I think that a short comparison of the use of these methods in the framework of perturbation theory is timely.

Loop ordered perturbative renormalization is the natural development of QED and has produced exceptionally successful phenomenological analyses in the framework of the Standard Model of Electro-Weak and Strong Interactions. Forgetting the problems related to infra-red divergences the construction of scattering amplitudes and operator matrix elements is based on the Feynman expansion with suitable subtraction prescriptions of the ultra-violet divergences. A systematic solution to the ultra-violet problem first described by Bogoliubov and Parasiuk, has found, after Hepp corrections,[4] a clear and handy form in Zimmermann's scheme subsequently extended in collaboration with Lowenstein to the massless case,[2] and by Breitenlohner and Maison to dimensional regularization.[5] The availabilty of this approach has led to many achievements such as a rigorous renormalized construction of gauge theories, systematic construction of renormalized operators, a clear and rigorous study of short distance physics.

Wilson's renormalization group was introduced as an alternative approach to Quantum Field Theory based on a systematic analysis of the scale transformation properties of Green functions. The natural framework is Euclidean field theory which can be related to a corresponding Minkowskian theory on the basis of Osterwalder-Schrader axioms.[1] The main goal consists in the construction of the Feynman-Kac functional integral. The most relevant application is the construction of gauge theories regularized on a lattice. The main purpose was and still is a non-perturbative construction of QCD and, in particular, the proof of confinement. On a lattice a scale transformation corresponds to the repeated replacement of the local fields with their averages over lattice cells. One studies the behavior of Feynman-Kac integral under these repeated substitutions. In the case of a theory built over a continuous manifold the analysis of scale transformation on the Feynman-Kac functional measure leads to a differential evolution equation for the measure.

In principle these evolution equations apply to the exact functional measure and do not rely on any Feynman graph expansion, however, until now, direct application of Wilson's approach to the construction of field theories beyond perturbation theory have been limited to special, however important, classes of models among which the most successful have been those involving only fermionic, and hence nilpotent, field variables. The construction of the Gross-Neveu model is the best known example.[6] In the general case one has to deal with an infinite sequence of equations that, in the case of bosonic variables, have no natural truncation. In some situation it is possible to justify the assumption of a measure remaining local after scale transformations;[7] this opens a further way toward non-perturbative results. However quite often the infinite sequence of evolution equations is truncated in a completely arbitrary way, often mimicking results that traditionally were obtained from naively simplified and truncated versions of the Schwinger-Dyson equations. The exact renormalized version of the Schwinger-Dyson equations has been studied in the early sixties by Symanzik and by Wu; the case of a scalar theory in four dimensions has been discussed by Johnson.[8] The analogy of this technique with Wilson's method should be better understood.

The application of the evolution equations to the construction of renormalized perturbation theory described by Polchinski in his thesis attracted new attention on Wilson's construction[9].[10] The essential reason for this interest lies in the major simplicity of the approach which is not directly based on a diagrammatic expansion. That is: the perturbative expansion of the functional measure leads to a series of terms each of which corresponds to a set of diagrams. Thus, even in a perturbative approach, the evolution equations deal with sets of diagrams, instead of dealing with single diagrams as the subtraction method does. Furthermore the differential nature of the evolution equations overcomes the problem of overlapping divergences. This, as shown by Hepp,[4] is the most difficult part of the Bogoliubov's renormalization project. In the renormalization group approach the overlapping divergences are disentangled by the cut-off derivative appearing in the evolution equation. This is just a pedagogical advantage, since one does not need anymore to have recourse to forests, however one should not underestimate a pedagogical advantage in a moment in which field theory is loosing part of the original interest being often presented as a special limit of a more general string "theory". On the other hand one should not consider Wilson-Polchinski method as an alternative computational method of renormalized amplitudes. Indeed the purpose of the short note is to prove that the pertubative solution to the evolution equations leads to a Zimmermann subtracted Euclidean field theory.

Taking into account the limits of this note we shall try to give a general idea of the reasons for this equivalence avoiding the formal aspects of a rigorous proof.[11]

2. The renormalization group evolution equations

With the aim described in the introduction we shall limit our discussion to the most simple situation considering an Euclidean scalar field theory in 4 dimensions.

Wilson's functional measure corresponds to an *Effective Interaction* which, when expanded into Feynman diagrams, is identified with the functional generator of connected amputated amplitudes built with the bare interaction and a doubly cut-off propagator, that is, with a propagator carrying an ultra-violet cut-off Λ_0 and an infra-red one Λ.

Wilson's equations describe the evolution of the measure with respect to Λ. The crucial part of the analysis consists in the proof that the Effective Interaction has a regular $\Lambda_0 \to \infty$, fixed Λ, limit. The final goal should be the study of the infra-red limit, i.e. $\Lambda \to 0$, which leads back to the renormalized (Schwinger) functions. However, fixing our mind on the ultra-violet problem, we limit our discussion to a pre-infra-red situation in which the infra-red cut-off Λ does not vanish. In this situation, if we restrict our discussion to perturbation theory, the role of the mass turns out to be of limited interest. On the other hand, inserting a mass into the propagator in perturbation theory, the $\Lambda \to 0$ limit becomes trivial. [a] Thus we do not pay particular attention to the $\Lambda \to 0$ limit and hence to the difference between Wilson's Effective Interaction and the generator of connected Green functions. This difference becomes relevant whenever there are infrared problems that we do not want to face.

Therefore we introduce the ultra-violet-infra-red cut-off Fourier transformed propagator:

$$\tilde{\hat{S}}(p) = \frac{e^{-\frac{p^2}{\Lambda_0^2}} - e^{-\frac{p^2}{\Lambda^2}}}{p^2} \tag{1}$$

and we define:

$$\Lambda^2 \frac{\partial}{\partial \Lambda^2} \tilde{\hat{S}}(p) \equiv \dot{\tilde{\hat{S}}}(p) = -\frac{e^{-\frac{p^2}{\Lambda^2}}}{\Lambda^2} . \tag{2}$$

Even if the best known version of the renormalization group evolution equation describes the Λ dependence of the Effective Interaction, for renormalization purposes it is convenient to consider the evolution equation of the Legendre transform of the Effective Interaction which is identified with the functional generator of the one-particle irreducible (1-P.I.) diagrams built with the bare interaction and the above propagator[12].[11] We call this new functional *1-P.I. Effective Action* and we label it with V_{Λ,Λ_0}.

The evolution equation of the 1-P.I. Effective Action can be easily deduced noticing that the Λ-derivative of each term of its expansion in Feynman diagrams only acts on propagators. If one selects and cuts a line into an one-particle irreducible diagram, what remains is an amputated connected

[a]If however one tries to have a look beyond perturbation theory one immediately encounters well known naturalness problems concerning the masses of scalars.

diagram consisting of a chain of 1-P.I. parts linked by single lines. Therefore the evolution equation can be represented as in the following figure:

$\Lambda\partial_\Lambda V_{\Lambda,\Lambda_0} \equiv \Lambda\partial_\Lambda$

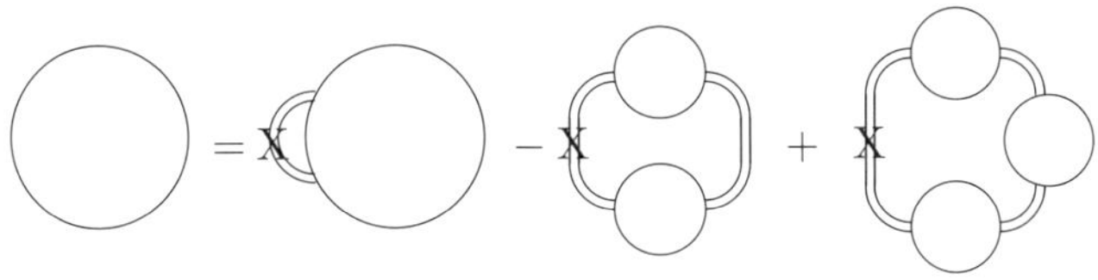

$$+ \cdots \equiv R_{\Lambda,\Lambda_0} \tag{3}$$

where double lines correspond to the propagator $\hat{S}$ and the crossed double one to $\dot{\hat{S}}$ while circles correspond to the 1-P.I. parts generated by V_{Λ,Λ_0}. The same equation in functional form appears as:

$$\Lambda^2 \frac{\partial}{\partial \Lambda^2} V_{\Lambda,\Lambda_0}[\phi] \equiv \dot{V}_{\Lambda,\Lambda_0}[\phi] = \frac{1}{2} Tr \left(\dot{\hat{S}} \frac{\delta^2 V_{\Lambda,\Lambda_0}}{\delta\phi^2} \sum_{n=0}^{\infty} (- * \hat{S} * \frac{\delta^2 V_{\Lambda,\Lambda_0}}{\delta\phi^2})^n \right)$$
$$\equiv \frac{1}{2} R_{\Lambda,\Lambda_0}[\phi] \; . \tag{4}$$

In the right-hand side of this equation $\dot{\hat{S}}$, $\hat{S}$ and $\delta^2 V_{\Lambda,\Lambda_0}/\delta\phi^2$ are multiplied as matrices and the traces of products are taken.

We translate this equation into a system of ordinary differential equations expanding:

$$V_{\Lambda,\Lambda_0}[\phi] = \sum_{n=0}^{\infty} \frac{1}{n!} \int \prod_{i=1}^{n} (dp_i \tilde{\phi}(p_i)) \delta(\sum_{j=1}^{n} p_j) V_n(p_1, \cdot\cdot, p_n, \Lambda, \Lambda_0)$$

and introducing an analogous expansion for $R_{\Lambda,\Lambda_0}[\phi]$. Notice that the coefficients $R_n(p_1, \cdot\cdot, p_n, \Lambda, \Lambda_0)$ of the field expansion of $R_{\Lambda,\Lambda_0}[\phi]$ are sums of series of terms corresponding to increasing numbers of 1-P.I. parts. Indeed this is apparent from Fig.(3). However, if we consider loop expanded quantities, the contribution of loop order ν to $R_n(p_1, \cdot\cdot, p_n, \Lambda, \Lambda_0)$ appears as a finite sum of terms built with the contribution of lower loop order of the coefficients $V_{n'}$ with $n' \leq n+2$. Thus, if V_2 vanishes at zero loop order, one never encounters infinite series[b].

Next step consists in translating this infinite system of differential equations into a corresponding system of integral equations accounting for the

[b]A mass term at zero loop order should be inserted into the propagator Eq. (1).

initial conditions of the evolution equation. In order to do this we need consistent bounds on the coefficients V_n and R_n. Using Eq.(4) it is not difficult to show[11] that, if up to loop order ν and uniformly in Λ_0, one has

$$\sup_p |\partial_p^k \, V_n(p_1, \cdot\cdot, p_n, \Lambda, \Lambda_0)| \leq \Lambda^{4-n-k} P_{n,k,\nu} \left(\log\left(\Lambda\right)\right)$$

a completely analogous bound holds true for $\sup_p |\partial_p^k \, R_n(p_1, \cdot\cdot, p_n, \Lambda, \Lambda_0)|$. Then the system of integral equations:

$$V_2(0,0,\Lambda,\Lambda_0) = \mu^2 + \int_{\Lambda_R}^{\Lambda} \frac{d\lambda}{\lambda} R_2(0,0,\lambda,\Lambda_0)$$

$$\partial_{p^2} V_2(p,-p,\Lambda,\Lambda_0)|_{p=0} = \zeta^2 + \int_{\Lambda_R}^{\Lambda} \frac{d\lambda}{\lambda} \partial_{p^2} R_2(p,-p,\lambda,\Lambda_0)|_{p=0}$$

$$V_4(0,\cdot\cdot,0,\Lambda,\Lambda_0) = g + \int_{\Lambda_R}^{\Lambda} \frac{d\lambda}{\lambda} R_4(0,\cdot\cdot,0,\lambda,\Lambda_0) \ ,$$

and, for $n + k > 4$,

$$\partial_p^k \, V_n(p_1, \cdot\cdot, p_n, \Lambda, \Lambda_0) = \int_{\Lambda_0}^{\Lambda} \frac{d\lambda}{\lambda} \, \partial_p^k \, R_n(p_1, \cdot\cdot, p_n, \lambda, \Lambda_0)$$

solves the evolution equations generating higher loop order terms in V_n satisfying analogous bounds. Now these bounds turn our to involve polynomials in $\log(\Lambda/\Lambda_R)$.

Furthermore both V_n and R_n have regular $\Lambda_0 \to \infty$ limits.

In this way one proves that the evolution equations produce a formally loop expanded 1-P.I. Wilson's Effective Action $V_R[\phi, \Lambda, \Lambda_R]$ which is defined as $\lim_{\Lambda_0 \to \infty} V_{\Lambda,\Lambda_0}[\phi]$ and whose field expansion coefficients satisfy the system of integral equations:

$$V_{R,2}(0,0,\Lambda,\Lambda_R) = \mu^2 + \int_{\Lambda_R}^{\Lambda} \frac{d\lambda}{\lambda} R_{R,2}(0,0,\lambda,\Lambda_R)$$

$$\partial_{p^2} V_{R,2}(p,-p,\Lambda,\Lambda_R)|_{p=0} = \zeta^2 + \int_{\Lambda_R}^{\Lambda} \frac{d\lambda}{\lambda} \partial_{p^2} R_{R,2}(p,-p,\lambda,\Lambda_R)|_{p=0}$$

$$V_{R,4}(0,\cdot\cdot,0,\Lambda,\Lambda_R) = g + \int_{\Lambda_R}^{\Lambda} \frac{d\lambda}{\lambda} R_{R,4}(0,\cdot\cdot,0,\lambda,\Lambda_R) \ , \qquad (5)$$

and, for $n + k > 4$,

$$\partial_p^k \, V_{R,n}(p_1, \cdot\cdot, p_n, \Lambda, \Lambda_R) = \int_{\infty}^{\Lambda} \frac{d\lambda}{\lambda} \, \partial_p^k \, R_{R,n}(p_1, \cdot\cdot, p_n, \lambda, \Lambda_R) \qquad (6)$$

where $R_{R,n}(p_1, \cdot\cdot, p_n, \Lambda, \Lambda_R) = \lim_{\Lambda_0 \to \infty} R_n(p_1, \cdot\cdot, p_n, \Lambda, \Lambda_0)$. It is apparent that the renormalized 1-P.I. Effective Action satisfies a differential evolution equation which is straightforwardly obtained from Fig.(3)

and Eq.(4) replacing V_{Λ,Λ_0} with V_R and the propagator in Eq.(1) with: $(1 - \exp(p^2/\Lambda^2))/p^2$.

Now we can specify the purpose of this note as follows: we want to show that the contribution of every single diagram to the solutions to the integral equations (5) and (6) and hence to the renormalized version of Fig.(3) and Eq.(4) corresponds to a suitably subtracted version of the Feynman amplitude associated with the diagram.

Notice that our set of integral equations ((5) and (6)) can be extended to the 1-P.I. Effective Action in the presence of local composite operators. Formally to every operator one couples an independent external field, whose dimension is obviously related to that of the operator. The evolution equations for the coefficients of the field-external-field expansion of the 1-P.I. Effective Action can be translated into integral equations accounting for initial conditions strictly analogous to Eqs.(5) and (6). It turns out[11] that the resulting renormalized composite operators directly correspond the Zimmermann's $N_\delta[P(\phi)]$ renowned operators.

3. Comparison with Zimmermann's subtraction approach

Here we come to the main goal of this note showing that in the $\Lambda_0 \to \infty$ limit an alternative construction of the iterative, loop expanded, solutions to the R.G. integral equations is given by an Euclidean variant of Zimmermann's (Lowenstein-Zimmermann) subtraction method. It is worth noticing that in many important instances the evolution equations are constrained by invariance conditions for the measure. The most frequently met are the Slavnov-Taylor-Ward identities. These conditions constrain the choice of the initial parameters. There are situation in which the constraints have no solution and hence one finds *anomalies*, the typical case is that of naive scale invariance. The analysis of invariance conditions is a crucial step of renormalization theory, we do not discuss it here since it is shown in the existing literature that this analysis follows the same lines in Wilson-Polchinski and subtraction approaches[11].[10]

In our simplified example the unsubtracted, and hence possibly divergent, Feynman integral corresponding to the diagram Γ contributing to the Schwinger function $S_n^{(m)}$ with an even number, n, of external legs and m loops, has the form:

$$S_\Gamma(p) = \int \frac{d^{4m}k}{(2\pi)^{4m}} I_\Gamma(p,k) \; ,$$

where $k \equiv k_1, ..., k_m$ is a basis of internal momenta of the diagram and

$p \equiv p_1, ..., p_{n-1}$ a basis of external momenta. $I_\Gamma(p,k)$ is built with the propagator:

$$\tilde{\hat{S}}(p) = \frac{1 - e^{-\frac{p^2}{\Lambda^2}}}{p^2} \tag{7}$$

and vertices

$$(\mu^2\phi^2 + \zeta^2(\partial\phi)^2)/2 \ , \quad g\phi^4/4! \ . \tag{8}$$

The subtraction procedure consists in replacing $I_\Gamma(p,k)$ with the renowned *forest formula*:

$$R_\Gamma(p,k) \equiv \mathcal{S}_\Gamma \sum_{F \in \mathcal{F}_\Gamma} \prod_{\gamma \in F} (-t^d_\gamma \mathcal{S}_\gamma) I_\Gamma(p.k) \tag{9}$$

where:

- $\mathcal{F}_\Gamma$ is the set of all forests of Γ
- $\mathcal{S}_\gamma$ defines the *momentum routing* in the sub-diagram γ
- t^d_γ takes the $\hat{p}^{(\gamma)}$ Taylor expansion of $I_\gamma(p,k)$ up to degree d_γ, the superficial divergence of γ,
- t^d_γ **replaces Λ with Λ_R in the propagators**.

Notice the analogy with Lowenstein-Zimmermann's[2] infrared subtraction scheme where an auxiliary parameter s is introduced, analogous to our Λ, and the ultra-violet subtraction is made at $s = 0$, in our example at $\Lambda = \Lambda_R$. We do not perform any infra-red subtraction that we should apply if we were interested in the $\Lambda \to 0$ mass-less limit.

Let us call $\mathcal{V}_\Lambda[\phi]$ the functional generator of the subtracted 1-P.I. Feynman amplitudes. The coefficient function $\mathcal{V}_n(p,\Lambda)$ of its field expansion appears as loop ordered formal series whose term of order ν is the sum of all the n-legs, ν-loops, subtracted 1-P.I. diagrams. We have to show that these coefficient functions satisfy the system of integral evolution equations (5) and (6).

The basic point that we have to show is that the Λ-derivative commutes with the subtraction operator as a consequence of the Λ-independence of the subtraction point. We start our analysis studying the Λ-derivative of a subtracted graph.

In order to do this let us take the Λ-derivative of a generic subtracted Feynman integral corresponding to a 1-P.I. diagram and hence contributing to $\mathcal{V}_\Lambda$. Due to the absolute convergence of the momentum integral we are allowed to commute this derivative with the internal momentum integration and hence we come to the k-momentum integral of $\partial_\Lambda R_\Gamma(p,k)$. As already done we notice that an un-subtracted Feynman integrand depends on Λ only through the propagators $\hat{S}$ and that the sub-diagram subtraction terms generated by the Taylor operators $t_\gamma^{d_\gamma}$ are Λ-independent since they are computed at $\Lambda = \Lambda_R$. Thus, in order to compute the Λ-derivative, we have single out in Eq.(9) the contributions of the propagators of un-subtracted sub-diagrams.

For a generic 1-P.I. diagram Γ we define:

$$R_\Gamma(p,k) = (1 - t_\Gamma^{d_\Gamma})\hat{R}_\Gamma(p,k) \tag{10}$$

where

$$\hat{R}_\Gamma(p,k) = \mathcal{S}_\Gamma \sum_{F \in \mathcal{F}'_\Gamma} \prod_{\gamma \in F} (-t_\gamma^d \mathcal{S}_\gamma) I_\Gamma(p.k) \tag{11}$$

and $\mathcal{F}'_\Gamma$ is the set of forests non containing Γ as an element. In other words, computing $\hat{R}_\Gamma(p,k)$ we exclude the subtraction of the whole diagram. The reason for this definition lies in the equation:

$$\partial_\Lambda R_\Gamma(p,k) = \partial_\Lambda \hat{R}_\Gamma(p,k) \ , \tag{12}$$

which means that, computing the Λ-derivative, one restricts the sum over the forests in Eq.(9) to $\mathcal{F}'_\Gamma$.

Now some more diagrammatic analysis is needed. For every forest F in $\mathcal{F}'_\Gamma$, we say that $\bar{\gamma} \in F$ is a maximal element of F if it is not contained into other elements of F. Then we call $\bar{F}$, *maximal sub-forest* of F, the set of maximal elements of F. Finally we label by $\bar{\mathcal{F}}'_\Gamma$ the set of maximal sub-forests in $\mathcal{F}'_\Gamma$. Notice that $\bar{\mathcal{F}}'_\Gamma$ coincides with the set of forests made of mutually disjoint sub diagrams of Γ.

A generic maximal sub-forest can be graphically represented as in the following figure:

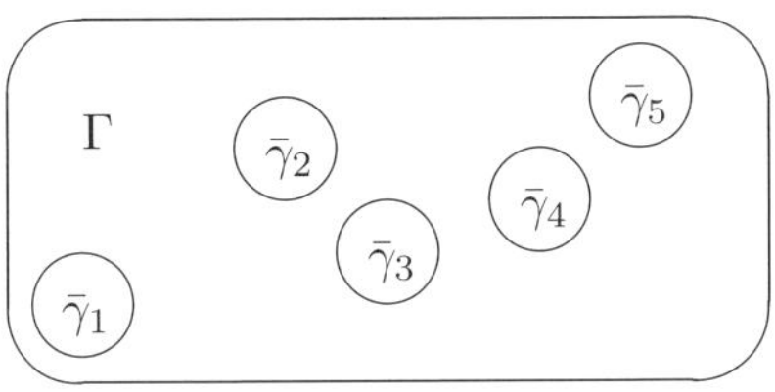

It is clear that any forest F in $\mathcal{F}'_\Gamma$ is equal to the union of forests $F_{\bar{\gamma}}$ contained in $\bar{\gamma}$ and including it as an element, for every $\bar{\gamma}$, element of the maximal sub-forest $\bar{F}$, that is:

$$F \equiv \cup_{\bar{\gamma} \in \bar{F}} F_{\bar{\gamma}}|_{\bar{\gamma} \in F_{\bar{\gamma}}} \ . \tag{13}$$

Therefore we can write Eq.(11) in the form:

$$\hat{R}_\Gamma(p,k) = \mathcal{S}_\Gamma \sum_{\bar{F}\in\bar{\mathcal{F}}'_\Gamma} \prod_{\bar{\gamma}\in\bar{F}} \sum_{F_{\bar{\gamma}}\in\mathcal{F}_{\bar{\gamma}},\bar{\gamma}\in F_{\bar{\gamma}}} \prod_{\gamma\in F_{\bar{\gamma}}} (-t^d_\gamma \mathcal{S}_\gamma) I_\Gamma(p.k) \ . \tag{14}$$

Given a maximal sub-forest $\bar{F}$ of Γ we define the *reduced diagram* $\Gamma/(\bigoplus_{\bar{\gamma}\in\bar{F}}\bar{\gamma})$ which is built with the lines and vertices of Γ not belonging to any element of $\bar{F}$ and of a further set of vertices corresponding to the elements $\bar{\gamma}$ of $\bar{F}$ shrunk to point vertices. The reduced diagram $\Gamma/(\bigoplus_{\bar{\gamma}\in\bar{F}}\bar{\gamma})$ is relevant to our discussion since the corresponding integrand identifies the part of $I_\Gamma(p.k)$ which is not concerned by the subtraction operation corresponding to the forest F. Indeed one can write:

$$\hat{R}_\Gamma(p,k) = \mathcal{S}_\Gamma \sum_{\bar{F}\in\bar{\mathcal{F}}'_\Gamma} \left[\prod_{\bar{\gamma}\in\bar{F}} ((-t^{d_{\bar{\gamma}}}_{\bar{\gamma}} \mathcal{S}_{\bar{\gamma}})\hat{R}_{\bar{\gamma}}(p,k))\right] I_{\Gamma/(\bigoplus_{\bar{\gamma}\in\bar{F}}\bar{\gamma})}(p,k) \tag{15}$$

This expression is identical to that associated with a diagram coinciding with the reduced diagram in which the vertices corresponding to the elements $\bar{\gamma}$ in $\bar{F}$ carry factors equal to $(-t^{d_{\bar{\gamma}}}_{\bar{\gamma}} \mathcal{S}_{\bar{\gamma}})\hat{R}_{\bar{\gamma}}(p,k)$. These factors, i.e. the brackets above, are, of course, Λ-independent. Therefore, inserting Eq.(15) into Eq.(12) one has:

$$\begin{aligned}
\Lambda^2\partial_{\Lambda^2} R_\Gamma(p,k) &= \mathcal{S}_\Gamma \sum_{\bar{F}\in\bar{\mathcal{F}}'_\Gamma} \left[\prod_{\bar{\gamma}\in\bar{F}} ((-t^{d_{\bar{\gamma}}}_{\bar{\gamma}} \mathcal{S}_{\bar{\gamma}})\hat{R}_{\bar{\gamma}}(p,k))\right] \Lambda^2\partial_{\Lambda^2} I_{\Gamma/(\bigoplus_{\bar{\gamma}\in\bar{F}}\bar{\gamma})}(p,k) \\
&= \mathcal{S}_\Gamma \sum_{\bar{F}\in\bar{\mathcal{F}}'_\Gamma} \left[\prod_{\bar{\gamma}\in\bar{F}} ((-t^{d_{\bar{\gamma}}}_{\bar{\gamma}} \mathcal{S}_{\bar{\gamma}})\hat{R}_{\bar{\gamma}}(p,k))\right] \\
&\quad \sum_{l\in L(\Gamma/(\prod_{\bar{\gamma}\in\bar{F}}\bar{\gamma}))} \dot{\hat{S}}(\hat{p}_l+\hat{k}_l) I_{\Gamma/(l\bigoplus_{\bar{\gamma}\in\bar{F}}\bar{\gamma})}(p,k)
\end{aligned}$$

where $\Gamma/(l\bigoplus_{\bar{\gamma}\in\bar{F}}\bar{\gamma})$ means the reduced diagram $\Gamma/(\bigoplus_{\bar{\gamma}\in\bar{F}}\bar{\gamma})$ deprived of the line l and we have used the fact that the Λ-dependence comes from the propagators.

Now we interchange the summation over the forests with that over the lines of Γ upon which the Λ-derivative acts. This is possible since every line l contributes to the above sum in correspondence with the forests F in $\mathcal{F}'_\Gamma$ whose elements do not contain it. If we extend the idea of forest to diagrams, such as Γ/l which are connected but not necessarily 1-P.I., the set of forests we are speaking of is $\mathcal{F}_{\Gamma/l}$ which, of course, is contained in

$\mathcal{F}'_\Gamma$. Thus we get:

$$\Lambda^2\partial_{\Lambda^2}R_\Gamma(p,k)=\mathcal{S}_\Gamma\sum_{l\in L(\Gamma)}\dot{\hat{S}}(\hat{p}_l+\hat{k}_l)\sum_{F\in\mathcal{F}_{\Gamma/l}}\prod_{\gamma\in F}(-t^d_\gamma\mathcal{S}_\gamma)I_{\Gamma/l}(p,k)\,, \tag{16}$$

Let us now consider the possibility of Γ/l not being 1-P.I.. The diagram Γ/l is however connected and it decomposes according to its skeleton structure into lines linking 1-P.I. parts. In the present situation, in which the diagram is obtained from a 1-P.I. diagram cutting the line l, Γ/l is either 1-P.I. or consists in a chain 1-P.I. sub-diagrams linked by lines. Therefore $I_{\Gamma/l}(p,k)$ factorizes into a product of line and 1-P.I. factors, one of the end points of the line l being attached to the first 1-P.I. sub-diagram of the chain, the other one to the last. Labelling these sub-diagrams by α_i , $i=0,..,n(\Gamma,l)$, where $n(\Gamma,l)$ is a non-negative integer, we can write:

$$I_{\Gamma/l}(p,k)=I_{\alpha_0}(p,k)\prod_{i=1}^{n(\Gamma,l)}\hat{S}(\hat{p}_i+\hat{k}_i))I_{\alpha_i}(p,k)\,. \tag{17}$$

If Γ/l is 1-P.I., the product above reduces to one.

Now a forest F in Γ/l appears as the union of, possibly trivial, forests in the above mentioned chain of 1-P.I. sub-diagrams, therefore the sum over the forests in $\mathcal{F}_{\Gamma/l}$ decomposes into the product of the sums over the forests in each sub-diagram α_i and hence we have:

$$\begin{aligned}\Lambda^2\partial_{\Lambda^2}R_\Gamma(p,k)&=\mathcal{S}_\Gamma\sum_{l\in L(\Gamma)}\dot{\hat{S}}(\hat{p}_l+\hat{k}_l)\sum_{F\in\mathcal{F}_{\Gamma/l}}\prod_{\gamma\in F}(-t^d_\gamma\mathcal{S}_\gamma)I_{\Gamma/l}(p,k)\\&=\mathcal{S}_\Gamma\sum_{l\in L(\Gamma)}\dot{\hat{S}}(\hat{p}_l+\hat{k}_l)\left[\mathcal{S}_{\alpha_0}\sum_{F\in\mathcal{F}_{\alpha_0}}\prod_{\gamma\in F}(-t^{d_\gamma}_\gamma\mathcal{S}_\gamma)I_{\alpha_0}(p,k)\right]\\&\prod_{i=1}^{n(\Gamma,l)}\left[\hat{S}(\hat{p}_i+\hat{k}_i))\left[\mathcal{S}_{\alpha_i}\sum_{F'\in\mathcal{F}_{\alpha_i}}\prod_{\gamma'\in F'}(-t^{d_{\gamma'}}_{\gamma'}\mathcal{S}_{\gamma'})I_{\alpha_i}(p,k)\right]\right]\\&=\mathcal{S}_\Gamma\sum_{l\in L(\Gamma)}\dot{\hat{S}}(\hat{p}_l+\hat{k}_l)R_{\alpha_0}(p,k)\prod_{i=1}^{n(\Gamma,l)}\left[\hat{S}(\hat{p}_i+\hat{k}_i))R_{\alpha_i}(p,k)\right].\end{aligned} \tag{18}$$

Now we consider how the Λ-derivative of a diagram must be subtracted in order to have absolute convergent internal momentum integrals. The basic remark is that the Λ-derivative only acts on lines giving (Eq.(2)) $-\exp((\hat{p}_l+\hat{k}_l)^2/\Lambda^2)/\Lambda^2$. Therefore we see that $\dot{\hat{S}}$ introduces a cut-off in the corresponding line momentum $(\hat{k}_l)$ and the needed subtraction formula must be limited to the forests in $\mathcal{F}_{\Gamma/l}$. Hence one gets back Eq.(16) and the commutativity of subtraction and Λ-derivative is proven.

Summing over all diagrams one also sums over all the possible values of $n(\Gamma, l)$ and it clearly appears that the structure of the rightmost term of Eq.(18) coincides with that of the right-hand side of the evolution equation of the Effective Action $V_R[\phi]$ and, of course, with that of its coefficient functions. Indeed one finds a sum over the chains of n 1-P.I. amplitudes linked by propagators $\hat{S}$ and closed by $\dot{\hat{S}}$.

It remains to verify the correct counting of diagrams. In other words until now we have shown that, computing the Λ-derivative of every 1-P.I. subtracted diagram, one gets a combination of subtracted diagrams with the structure appearing in Fig.(3). What remains is a purely combinatorial problem, that is to verify that computing the Λ-derivative of $\mathcal{V}_\Lambda$, that is summing all diagrams together, one gets an expression in which all the expected diagrams appear with the expected combinatorial factor. This is just the consequence of the discussed commutativity of subtraction and Λ-derivative. Indeed the fact that before subtraction all the expected diagrams appear with the right factors is proven by a straightforward application of the functional method. At the formal level, disregarding divergences, the functional generator of Feynman diagrams Z is perfectly well defined, the generator of connected diagrams is $\ln Z$ and that of 1-P.I. diagrams is the Legendre transform of $\ln Z$. Now it is easy to show[11] that Eq.(4) is satisfied by the formal graph expansion under the hypothesis that the derivative only acts on lines. This guarantees the correct counting of diagrams and completes our proof.

In order to give a significant example let us consider the three line, two leg, diagram shown in Fig.(19):

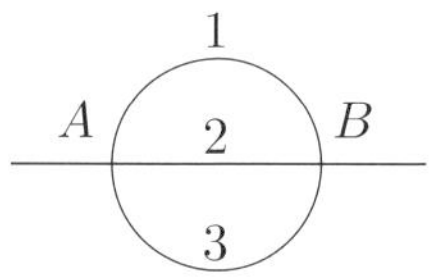

(19)

This diagram seems to violate what just claimed, indeed it contains three indistinguishable internal lines, and hence its Λ-derivative gives three identical contributions in which $\dot{\hat{S}}$ is linked to a single diagram with two identical lines. On the contrary, in a diagrammatic expansion of Fig.(3) and Eq.(4) this diagram should appear only once. This is however a wrong argument since it forgets the combinatorial factors of the diagrams. A diagram with N sets of n_i, $i = 1, ..., N$, indistinguishable lines carries a combinatorial factor equal to $1/(\prod_{i=1}^{N} n_i!)$ that is 1/6 in the example. Combining the three identical contributions from the three lines together we get the resulting contribution to the evolution equation with weight 1/2 which is exactly the combinatorial factor of the corresponding diagram with two identical lines.

In conclusion we have shown that, applying a slightly modified subtraction method to the Feynman diagrams built with the propagator $\hat{S}$ given in Eq.(7), and possibly with its spinor, or gauge field variants, yields to a diagrammatic construction of $\mathcal{V}_\Lambda[\phi]$ solving the R.G. evolution equation (4).

However we want also to show that the field expansion coefficients of $\mathcal{V}_\Lambda[\phi]$ satisfy Eqs.(5) and (6) with the initial conditions at $\Lambda = \Lambda_R$ appearing in Eq.(5), and furthermore that the limit $\Lambda \to \infty$ of $\partial_p^k\ \mathcal{V}_{\Lambda,n}$ for $n + k > 4$ vanishes.

It is apparent that $\mathcal{V}_{\Lambda,2}$ and $\mathcal{V}_{\Lambda,4}$ satisfy Eqs.(5). Indeed $\mathcal{V}_{\Lambda,2}$ is the sum of two leg proper diagrams which, with the exception of the trivial diagrams generated by the first two vertices in Eq.(8), are superficially divergent and hence subtracted to zero at $p = 0$ and $\Lambda = \Lambda_R$ with their first derivative in p^2. Furthermore $\mathcal{V}_{\Lambda,4}$ is the sum of four leg proper diagrams which, with the exception of the trivial diagram generated by the third vertex in Eq.(8), are superficially divergent and hence subtracted to zero at $p = 0$ and $\Lambda = \Lambda_R$.

Concerning the derivatives of the coefficients $\partial_p^k\ \mathcal{V}_{\Lambda,n}$ for $n+k > 4$, they only receive contributions from superficially convergent diagrams which are easily seen to vanish in the $\Lambda \to \infty$ limit using the inequality:

$$(1 - \exp(-p^2/\Lambda^2))/p^2 \leq 2/(p^2 + \Lambda^2), \tag{20}$$

and pure scale arguments.

Therefore we conclude that the construction of the Effective Action $\mathcal{V}_\Lambda[\phi]$ by the above defined subtraction method leads to a solution of Wilson-Polchinski evolution equation satisfying the boundary conditions characterizing Wilson's construction, thus it leads to the same functional:

$$\mathcal{V}_\Lambda[\phi] \equiv V_R[\phi, \Lambda, \Lambda_R]\ .$$

4. Conclusions

In conclusion, comparing the Wilson-Polchinski renormalization group and the BPHZ subtraction approach one sees that in both cases one is dealing with an infinity of quantities related by an infinity of equations and hence the chosen ordering is a crucial step of the construction procedure.

The subtraction approach deals with one diagram at a time and the physical amplitudes appear as formal expansions into subtracted diagrams which must be ordered in some way. The loop ordering is the typical choice.

The R.G. integral equations (5) and (6) for the coefficient functions of the field expanded 1-P.I. effective action are not strictly related to diagrams, hence a wider class of recursive construction is *in principle* open. However the right-hand sides of the evolution equations appear as the sum of series which are infinite due to the presence of chains of two-point insertions which in principle can be summed. This is particularly critical in the scalar field case due to the quadratic divergence of the mass terms.

The set of evolution equations for the coefficient functions is infinite and open, in the sense that it does not contain any closed finite sub-set, that is, any finite sub-set of equations involving a finite number of coefficient functions. Indeed the evolution equation of the coefficient $V_{R,n}$ involves $V_{R,n+2}$. Thus, in order to build a solution, one must truncate in some way the sequence of the $V_{R,n}$ evolution equations.

We have limited our study to the loop ordered perturbative expansion in which the sequence of evolution equations appears closed at any order. This has allowed us to study the details of the resulting amplitudes proving that their expansion into diagrams coincides with that generated by the subtraction method with a suitable, however natural, choice of the subtraction prescriptions.

With the aim of simplifying our presentation we have also limited our discussion to the simplest scalar model disregarding invariance properties and possible infra-red singularities, thus, in a sense, remaining far apart from the physical applications.

Our hope is that the present discussion could further clarify the relations among different construction techniques of Quantum Field Theory confirming the central role of Zimmermann's work.

References

1. A general reference on Quantum Field Theory is:
   R. Haag, *Local Quantum Physics.* (Springer-Verlag, Berlin 1992).
2. For a general review and an exhaustive list of references on the BPHZ method see:
   J. H. Lowenstein, *BPHZ renormalization* in *Renormalization Theory. Proceedings of the NATO Advanced Study Institute held at the International School of Mathematical Physics at the 'Ettore Majorana' Center for Scientific Culture in Erice (Italy), G. Velo and A. Wightman Eds.* - (D.Reidel Publishing Company, Boston 1976) pg. 95.
   W. Zimmermann, *The power counting theorem for Feynman integrals with massless propagators*, ibid. pg. 171.
3. A general reference is:
   K. Wilson and J. Kogut, *Phys. Reports* **12**, 75 (1974).
4. K. Hepp, *Commun.Math.Phys.* **2**, 301 (1966).
5. P. Breitenlohner and D. Maison, *Commun. Math. Phys.* **52**, 11 (1977), ibid. pg. 39, ibid pg. 55.
6. K. Gawedzki, *Commun. Math. Phys.* **102**, 1 (1985).
7. T. R. Morris, *Phys.Lett.B* **329**, 241 (1994).
8. R. W. Johnson, *J.M.P.* **11**, 2161 (1970).
9. J. Polchinski, *Nucl. Phys. B* **231**, 269 (1984).
   G. Gallavotti, *Rev. Mod. Phys.* **57**, 471(1985).
10. An account of the Wilson-Polchinski approach with application to gauge theories is given by:
    C. Becchi *On the construction of renormalized quantum field theory using renormalization group techniques.* In *Elementary Particles, Quantum Fields and Statistical Mechanics. Seminario Nazionale di Fisica Teorica* M. Bonini, G. Marchesini, E. Onofri Eds. Parma 1993 (hep-th/9607188)
11. A more detailed analysis of the same subject is given in:
    C. Becchi, in http://www.ge.infn.it/∼becchi/prague-2007.pdf
12. M. Bonini et al., *Nucl.Phys.B* **409**, 441 (1993).

A NEW LOOK AT THE HIGGS-KIBBLE MODEL

OTHMAR STEINMANN*

Fakultät für Physik, Universität Bielefeld, 33501 Bielefeld, Germany
**E-mail: stema@physik.uni-bielefeld.de*

An elementary perturbative method of handling the Higgs-Kibble models and deriving their relevant properties, is described. It is based on Wightman field theory and avoids some of the mathematical weaknesses of the standard treatments. The method is exemplified by the abelian case. Its extension to the non-abelian gauge group SU_2 is shortly discussed in the last section.

Keywords: Gauge theories; Spontaneous symmetry breaking

1. Introduction

The spontaneous breaking of gauge invariance as described by the Higgs-Kibble model (henceforth HKM) is an essential ingredient of the electroweak part of the standard model of elementary particle physics. In the present work we will report on a new, rather elementary, method of deriving the properties of the model, in particular its renormalizability (or lack thereof, see Sect.6).

Our method is entirely perturbative, it consists predominantly in studying the properties of so-called 'sector graphs', a simple generalization of Feynman graphs. But the corresponding graph rules are derived in an unconventional way. We do not use path integrals, a not entirely convincing method because of the lack of a solid mathematical underpinning. Nor do we use the canonical formalism with its own weak points, like the dubious status of the canonical commutation relations on account of the non-existence of interacting fields at a sharp time, and the need for introducing and handling constraints. Instead we work with an adaptation of the method introduced in[6] for QED, where many details are found beyond what can be reported here.

We will concentrate on the case of the abelian HKM. The extension of our method, and of its results, to the non-abelian case will, however, be briefly described in the last section. Also, we will work throughout at a formal, non-renormalized level, only getting as far as obtaining the power-counting behavior necessary for establishing renormalizability. This sticking to non-renormalized expressions is not as bad as it sounds. We propose that the theory be renormalized by Zimmermann's method (known as BPHZ, see[9]), which consists in subtracting not integrals, but the integrands of the Feynman graphs or, in our case, the sector graphs. And the cancellations between graphs that we need to establish for obtaining renormalizability, also happen for the integrands. Therefore we need only talk about the well defined integrands, and the divergence (before renormalization) of the integration over them need not unduly bother us.

2. The Model

Let us start with a brief reminder of the definition of the HKM.[a] The abelian HKM is a relativistic field theory containing a complex scalar field $\Phi(x)$ and a real vector field $A_\mu(x)$. Its dynamics is specified by the Lagrangian

$$\begin{aligned} L = & -\frac{1}{4} F_{\alpha\beta} F^{\alpha\beta} + (\partial_\alpha - igA_\alpha)\,\Phi^*\,(\partial^\alpha + igA^\alpha)\,\Phi \\ & + \mu^2\,\Phi^*\Phi - \lambda\,(\Phi^*\Phi)^2 \end{aligned} \tag{1}$$

with

$$F_{\alpha\beta}(x) = \partial_\alpha A_\beta(x) - \partial_\beta A_\alpha(x). \tag{2}$$

g, λ, μ, are positive real numbers. An important feature of this Lagrangian is the 'wrong' sign of the mass term $\mu^2\Phi^*\Phi$. L is invariant under the gauge transformation

$$\Phi(x) \Rightarrow \exp[i\,g\,\vartheta(x)]\,\Phi(x), \quad A_\mu(x) \Rightarrow A_\mu(x) - \partial_\mu\vartheta(x) \tag{3}$$

for real functions ϑ.

Because of the unconventional mass term, the field equations derived from L possess the non-trivial classical solution of lowest energy

$$\Phi(x) = \Phi^*(x) = \frac{v}{\sqrt{2}}\,, \qquad A_\mu(x) = 0 \tag{4}$$

with

$$v = \mu/\sqrt{\lambda} > 0\,. \tag{5}$$

Other solutions of the same lowest energy are generated from (4) by applying gauge transformations (3). But they are of no concern to us.

[a]The standard lore about spontaneous symmetry breaking can be found e.g. in,[38]

Our perturbative quantum solution consists essentially in a quantum expansion around the real solution (4). We make the ansatz

$$\Phi(x) = \frac{1}{\sqrt{2}} \left(v + R(x) + i\, I(x)\right), \tag{6}$$

where R and I are two real fields. Henceforth we treat A_μ, R, I, as the fundamental fields of the model, while Φ is forgotten. With these new fields the solution (4) takes the trivial form

$$A_\mu = R = I = 0 \,. \tag{7}$$

The gauge transformation (3) can be transcribed into the new fields. We will not write the result down since we are not going to use it, apart from the important fact that $F_{\alpha\beta}$ and the 'Higgs field'

$$\Psi(x) = R(x) + \frac{1}{2v} \left[R^2(x) + I^2(x)\right] \tag{8}$$

are gauge invariant.

The Lagrangian (1) can also be transcribed into the new fields. It takes the form

$$L = L_2 + L_3 + L_4 \ , \tag{9}$$

where L_i collects the terms of order i in the fields. A constant term L_0 has been dropped as being immaterial. Furthermore, we replace λ, μ, as parameters of the theory by

$$m = g\,v = \frac{g\,\mu}{\sqrt{\lambda}}, \qquad M = \sqrt{2}\,\mu \ , \tag{10}$$

which denote the masses of the gauge boson and the Higgs particle respectively. They are therefore measurable quantities (barring the need for renormalization), and they will as usual be kept fixed. Perturbation theory amounts then to a power series expansion in the remaining coupling constant g. The L_i read

$$\begin{aligned} L_2 = &-\frac{1}{4} F_{\alpha\beta} F^{\alpha\beta} + \frac{m^2}{2} A_\alpha A^\alpha + m\, A^\alpha \partial_\alpha I \\ &+\frac{1}{2} (\partial_\alpha R\, \partial^\alpha R - M^2 R^2) + \frac{1}{2}\, \partial_\alpha I\, \partial^\alpha I \ , \end{aligned} \tag{11}$$

$$\begin{aligned} L_3 = &-g\, A^\alpha I \partial_\alpha R + g\, A^\alpha \partial_\alpha I\, R + g\, m\, A_\alpha A^\alpha R \\ &- \frac{gM^2}{2m} R^3 - \frac{gM^2}{2m} R\, I^2 \ , \end{aligned} \tag{12}$$

$$L_4 = \frac{1}{2}\, g^2 A_\alpha A^\alpha\, R^2 + \frac{1}{2}\, g^2 A_\alpha A^\alpha\, I^2 \\ -\frac{g^2M^2}{8m^2}\, R^4 - \frac{g^2M^2}{8m^2}\, I^4 - \frac{g^2M^2}{4m^2}\, R^2 I^2\ . \tag{13}$$

L_2 will be responsible for the propagators of our graph rules, $L_{int} = L_3 + L_4$ for the vertices. The Higgs field takes the form

$$\Psi(x) = R(x) + \frac{g}{2m}\,[R^2(x) + I^2(x)]\ . \tag{14}$$

In our method the dynamics is embodied in the field equations rather than in the Lagrangian. They take the form

$$(\Box + m^2)A^\mu - \partial^\mu\partial_\nu A^\nu + m\,\partial^\mu I = -\frac{\delta L_{int}}{\delta A_\mu} =: \mathcal{R}^\mu(x)\ , \tag{15}$$

$$-\Box I - m\,\partial_\nu A^\nu = -\frac{\delta L_{int}}{\delta I} =: \mathcal{R}_I(x)\ , \tag{16}$$

$$-(\Box + M^2)\, R = -\frac{\delta L_{int}}{\delta R} =: \mathcal{R}_R(x)\ . \tag{17}$$

As a consequence of the gauge freedom of the theory we note the following fact. Applying the derivation ∂_μ to the left-hand side of (15) we obtain the left-hand side of (16), up to a constant factor. The equations (15)–(17) can therefore possess solutions only if the *consistency condition*

$$\mathcal{F} := \partial_\mu \mathcal{R}^\mu + m\,\mathcal{R}_I = 0 \tag{18}$$

is satisfied. That this condition is satisfied in our case is essentially a consequence of the field equations having been derived from a Lagrangian. It must, however, be noted that in an explicit verification the field equations must be used. This verification runs as follows. As contribution of L_3 to $\mathcal{F}$ we find

$$\mathcal{F}_3 = g\, I\, (\Box + M^2)R - g\, R\, (\Box I + m\,\partial_\mu A^\mu)\ . \tag{19}$$

Using the field equations (16) and (17) this becomes a polynomial of order 3 in the fields which exactly cancels the L_4-contribution

$$\mathcal{F}_4 = -g^2R^2\partial_\mu A^\mu - 2\,g^2R\,\partial_\mu R\, A^\mu - g^2I^2\partial_\mu A^\mu - 2g^2I\partial_\mu I\, A^\mu \\ -m\,g^2I\, A_\mu A^\mu + \frac{g^2M^2}{2m}\, I^3 + \frac{g^2M^2}{2m}\, R^2 I\ . \tag{20}$$

This looks at first like a consistency check rather than a proof. vIt is, however, perfectly acceptable as a proof in perturbation theory.

We will not endeavor to give a general definition of what we understand under a particular gauge of this model. But the following statement is essential. *A quantum field theory claiming to be the HKM in a particular gauge must satisfy the field equations (15)–(17).* In the following section a particular class of gauges will be constructed.

3. Wightman Gauges

Under 'Wightman gauges' we understand a class of quantum field theories solving the field equations (15)–(17), and moreover satisfying the Wightman axioms (see[7]), i.e. Poincaré covariance, locality, spectral condition, existence of a vacuum, and the cluster property, with the possible exception of positivity. This last condition can in general not be expected to hold in a gauge theory. Two special cases of Wightman gauges will be of particular interest to us. The first is the 'unitary' or 'physical' gauge, which allows to specify the physical content of the theory. And the second is the 'renormalization' gauge, which is particularly suited for establishing the renormalizability of the physically relevant part of the model.

Our method consists essentially in a recursive solution of the field equations. But the fundamental objects of the approach are the Wightman functions (W-functions), not the field operators themselves, and also not the Green's functions of the conventional methods. The W-functions are the vacuum expectation values of ordinary (not time ordered) products of field operators. According to Wightman's reconstruction theorem[7] the theory is fully determined by these W-functions.[b] The field equations applied to any factor in a W-function produce a set of differential equations for these functions. And this set of differential equations we solve recursively.

The resulting expression for a given function $(\Omega, \varphi_1(x_1) \cdots \varphi_n(x_n)\,\Omega)$, φ_i any of the fundamental fields A_μ, I, R, in a given order g^σ of perturbation theory can be written as a sum over generalized Feynman graphs called 'sector graphs'. A sector graph looks at first just like an ordinary Feynman graph not containing any vacuum-vacuum subgraphs[c]. But its vertices are then partitioned into non-overlapping subsets called 'sectors', in such a way that each sector contains at most one external point corresponding to one of the fields in W. Lines connecting vertices (including the external points) in the same sector belong to this sector and are called 'sector lines'. Lines connecting points in different sectors are called 'cross lines'. The sectors

[b] Positivity of the scalar product is not necessary for the validity of the reconstruction theorem (see Sect. 4.2 of[6]).

[c] These subgraphs do not occur in our formulation because we work in the Heisenberg picture. The vacuum graphs are an artifact of the interaction picture.

can be of two types, T^+ or T^-. They are numbered such that the sector containing the external point belonging to φ_i carries the number i. It is convenient to alternate the corresponding sectors: sectors with an odd number are T^-, those with an even number T^+, or vice versa. In this case there occur no sectors without external points.

The internal vertices correspond as usual to the terms in L_{int} as listed in (12), (13). Their vertex factors are also the conventional ones in T^+ sectors, their complex conjugates in T^- sectors. E.g. the last term in (12) produces a vertex with one R-line and two I-lines joining, and with the vertex factor $\mp i\, g\, m^{-1} M^2$ in $T^\pm$ sectors. Note that the L_4-vertices are of second order in g. A cross line joining the vertex with variable u in sector i to the vertex v in sector j, $j > i$, carries the 'cross propagator'

$$w_{ab}(u - v) = \langle \varphi_a(u), \varphi_b(v) \rangle_0 \ , \tag{21}$$

where $\langle \varphi_a\, \varphi_b \rangle_0$ is a free 2-point function (to be specified below), and the indices a, b, signify the field types of the ends of the line in question. A sector line connecting the vertices u and v in a $T^\pm$ sector carries as propagator the time ordered or anti-time ordered function $\tau^\pm_{ab}(u - v)$ corresponding to the w_{ab} of (21).

With the rules given as yet there holds the *Ostendorf theorem*,[4],[5],[6] stating that the so defined functions W_σ satisfy all Wightman properties with the possible exception of positivity. Hence these rules define a, slightly generalized, Wightman theory.

But we still must satisfy the requirement that these W solve the interacting field equations (15)–(17). This problem is easier to handle in p-space. Therefore we will from now on mainly work in this space, with the Fourier transforms of (15)–(17). That these equations are satisfied in 0th order in g is guaranteed by the condition that the w_{ab} solve the free field equations. For the following we need to know the w_{ab} more explicitly. In p-space we have

$$\langle \varphi_a(p)\, \varphi_b(q) \rangle_0 = w_{ab}(p)\, \delta^4(p + q) \ , \tag{22}$$

where this new w_{ab} is the Fourier transform of the w_{ab} in (21). The most general solution of the free field equations satisfying all Wightman properties, in particular covariance and locality, is easily found to be

$$\left.\begin{aligned} w_{\mu\nu}(p) &= -\omega\left(g_{\mu\nu} - \tfrac{p_\mu p_\nu}{m^2}\right)\delta_+^m(p) + \tfrac{1}{m}\,p_\mu p_\nu\,T(p) \\ w_{II}(p) &= m\,T(p) \\ w_{\mu I}(p) &= -w_{I\mu}(p) = i\,p_\mu\,T(p) \\ w_{RR}(p) &= \alpha\,\delta_+^M(p) \\ w_{RI}(p) &= \beta\,\delta_+^M(p)\,, \qquad w_{R\mu}(p) = -i\,\tfrac{\beta}{m}\,p_\mu\,\delta_+^M(p)\,, \end{aligned}\right\} \tag{23}$$

where $\delta_+^m(p) = \theta(p_0)\,\delta(p^2 - m^2)$ is the Dirac measure for the positive mass shell. α, β, ω, are as yet undetermined real constants, $T(p) = \theta(p_0)\,T'(p^2)$ is an arbitrary real invariant function with support in the forward light cone.

The corresponding (anti-)time ordered functions $\tau^\pm(p)$ serving as sector propagators are then also uniquely fixed, provided we restrict ourselves to T' which tend to 0 for $p^2 \to \infty$, and that we demand that $\tau^\pm$ should increase for $p \to \infty$ as weakly as possible. It turns out that the resulting W-functions satisfy the interacting field equations, if the $\tau^\pm$ are propagators in the original sense of the word used in the theory of differential equations. In p-space this means the following. We write the p-space form of the field equations (15)–(17) in matrix notation as

$$C(p)\,\varphi(p) = \mathcal{R}(p)\ . \tag{24}$$

Here C is the 6×6 coefficient matrix

$$C = \begin{pmatrix} -(p^2 - m^2)\,\delta_\nu^\mu + p^\mu p_\nu & -i\,m\,p^\mu & 0 \\ i\,m\,p_\nu & p^2 & 0 \\ 0 & 0 & p^2 - M^2 \end{pmatrix}. \tag{25}$$

The lines are indexed by μ, I, R, the rows by ν, I, R, where μ and ν run over the values $0, \cdots, 3$. φ is a 6-vector with components $(A^\nu,\ I,\ R)$, $\mathcal{R}$ a 6-vector with components $(\mathcal{R}^\mu,\ \mathcal{R}_I,\ \mathcal{R}_R)$. We call the 6×6 matrix $\mathcal{P}(p)$ a *propagator matrix* if

$$C\,\mathcal{P}\,V = V \tag{26}$$

holds for all 6-vectors $V(p)$ satisfying the consistency condition (18):

$$-i\,p_\mu V^\mu + m\,V_I = 0\ . \tag{27}$$

Then

$$\varphi = \mathcal{P}\,\mathcal{R} \tag{28}$$

solves (24). Remember that the I-line of C is a linear combination of the μ-lines. Hence C is not invertible and $\mathcal{P}$ cannot be defined as its inverse. Therefore the restriction (27) is necessary.

The sum over our sector graphs solves the field equations of the HKM if

$$\mathcal{P}^{\pm}_{ab}(p) = \mp 2\pi i \, \tau^{\pm}_{ab}(p) \tag{29}$$

constitute a propagator matrix. This is seen by applying the field equations to the propagators of the external graph lines, using that the internal ends of these lines correspond to $\mathcal{R}$ vertices (see,[6] Sect. 9.4, for the QED analogue). External cross propagators do not contribute because they solve the free field equations. It turns out that condition (29) fixes two of the free constants in (23) to be

$$\omega = \alpha = 1 \; , \tag{30}$$

while β and the function $T(p)$ are still free.

From our rules for calculating W-functions we can also obtain the rules for the fully time ordered functions. At our present formal, non-renormalized, level this is simply done by using the formal definition of time ordering with the help of step functions. The result is a representation as a sum of graphs with only one T^+ sector containing all external points. The corresponding graph rules are simply the standard Feynman rules. That the Green's functions thus defined are indeed the time ordered functions of a field theory is of course essential for the applicability of the LSZ reduction formula for the calculation of the S-matrix.

We will also have occasion to consider functions of the form

$$\Big(\Omega, \, T^-\big(\varphi_1(x_1)\cdots\varphi_n(x_n)\big)\, T^+\big(\psi_1(y_1)\cdots\psi_m(y_m)\big)\,\Omega\Big) \; ,$$

where φ_i or ψ_j stands for any of our fields. These are given by 2-sector graphs with a T^- sector containing all external x_i points and a T^+ sector containing all y_j points.

4. The Unitary Gauge

The unitary gauge, or U-gauge, is defined as the special Wightman gauge obtained by the choice

$$\beta = T(p) = 0 \; . \tag{31}$$

In this gauge we have

$$I = 0 \; , \tag{32}$$

it is simply the gauge specified by the 'gauge condition' (32)[d]. Hence we are left only with the fields R and A_μ. The surviving non-vanishing cross propagators are

$$w_{\mu\nu}(p) = -\,(g_{\mu\nu} - m^{-2}p_\mu p_\nu)\,\delta_+^m(p)\ , \quad w_{RR}(p) = \delta_+^M(p)\ , \tag{33}$$

and the sector propagators are

$$\begin{aligned} \tau_{\mu\nu}^{\pm}(p) &= \mp(i/2\pi)\,(g_{\mu\nu} - m^{-2}\,p_\mu p_\nu)\,(p^2 - m^2 \pm i\epsilon)^{-1} \\ \tau_{RR}^{\pm}(p) &= \pm(i/2\pi)\,(p^2 - M^2 \pm i\epsilon)^{-1}\ . \end{aligned} \tag{34}$$

The special interest of this gauge rests on the fact that it might also be called the 'physical gauge'. The physically relevant objects of a quantum field theory are the observables and the physical states.[e] In an experiment we usually measure expectation values of observables in physical states (meaning states that can actually be prepared in a laboratory). The physical content of a gauge theory must be gauge independent. For the observables this implies that they must be gauge invariant. What it means for states is less easy to characterize. But in the HKM the state space $\mathcal{V}_U$ of the U-gauge is the obvious candidate for the role as physical state space. This claim rests on two facts. First, the cross propagators are positive. For the $w_{\mu\nu}$ this means more exactly that they form a positive matrix. This implies that our graph rules define on $\mathcal{V}_U$ a positive scalar product[f], a necessary requirement for a physical state space.

The second vital point is the following. At first, $\mathcal{V}_U$ is generated from the vacuum state Ω by applying to it polynomials in the fields R, A_μ, properly integrated over sufficiently smooth test functions. But it turns out that the same state space is also created out of Ω by applying polynomials in the gauge invariant fields $F_{\alpha\beta}$ and Ψ. This is so because A_μ and R can be expressed as functions of $F_{\alpha\beta}$ and Ψ. We see this as follows. The definition (14) of Ψ becomes in the U-gauge

$$\Psi = R + \frac{g}{2m}\,R^2\ . \tag{35}$$

[d]The other W-gauges cannot be characterized in this simple way.
[e]The widely held opinion that the physical content of the theory is fully described by its S-matrix is not tenable. The S-matrix relates states at positive infinite times to states at negative infinite time. But we always measure at finite times. Therefore the S-matrix is, in fact, not measurable.
[f]A formal power series $Q(g)$ is said to be positive if there exist formal power series $S_i(g)$ such that $Q(g) = \sum_i S_i(g)^*\,S_i(g)$.

This equation can in principle be solved for R. Of course, square roots of operators are not easy to deal with. But we work in perturbation theory, and here there is no problem. Expand R in a power series:

$$R = \sum_{\sigma=0}^{\infty} R_\sigma g^\sigma \ , \tag{36}$$

and similarly for Ψ, and insert these expansions into (35). We find in zeroth order

$$R_0 = \Psi_0 \ ,$$

in first order

$$R_1 = \Psi_1 - (2m)^{-1} {R_0}^2 = \Psi_1 - (2m)^{-1} {\Psi_0}^2 \ ,$$

and so on. The fact that for increasing σ R_σ becomes a polynomial in Ψ_ϱ, $\varrho \leq \sigma$, of indefinitely increasing order need not worry us, because this expansion will never be used explicitly. Next, from the definition of $F_{\alpha\beta}$ and the field equation (15) we obtain

$$\partial^\alpha F_{\alpha\beta} = \Box A_\beta - \partial_\beta \partial^\alpha A_\alpha = -m^2 A_\beta + \mathcal{R}_\beta(A_\mu, R) \ , \tag{37}$$

or its Fourier transform, hence

$$m^2 A_\beta = -\partial^\alpha F_{\alpha\beta} + \mathcal{R}_\beta(A_\mu, R) \ . \tag{38}$$

Since $\mathcal{R}_\beta$ contains an explicit factor g, this equation allows again an iterative expansion of A_μ in polynomials of $F_{\mu\nu}$ and Ψ.

Hence $\mathcal{V}_U$ has an explicitly gauge invariant structure, which fact justifies the claim that it is the physical state space of the HKM.

In this way we seem to have arrived at a nice, clean, identification of the physical content of the HKM. There is, however, a fly in the ointment. The $p_\mu p_\nu$ term in (34) has a bad behavior at large p, leading to non-renormalizability of the theory in the simple power counting sense. In increasing orders of perturbation theory the individual graphs will have an increasingly bad ultraviolet behavior. The claim that the theory is nevertheless renormalizable amounts to claiming that these bad UV contributions in individual graphs cancel in the sum of all graphs contributing to a specific W-function (or time ordered function) in a given order σ of perturbation theory.

The standard way of handling this problem consists in adding a so-called 'gauge fixing term'

$$-\frac{1}{2\,\alpha}\,(\partial_\mu A^\mu)^2 \tag{39}$$

to the original Lagrangian of the model. The theory thus obtained is renormalizable in the sense of power counting. But calling (39) a gauge fixing term is highly misleading. The amended α-Lagrangian does by no means describe a particular gauge of the HKM. It defines a new, different theory, which does *not* solve the field equations of the HKM. Hence its renormalizability is of no use to our problem, unless it can be established to be in some way physically equivalent to the HKM, in particular to its U-gauge formulation. This necessity does not quite find sufficient attention in the literature.

In any case, if the claimed cancellations between graphs really happen, this ought to be provable inside the HKM. This is the task that we now turn to. There seems to be no easy way to achieve this purpose in the U-gauge. Therefore we introduce in the next section another Wightman gauge better suited to the task.

5. Renormalizability

Particularly suited for our purpose is the *R-gauge* (for 'renormalization gauge') specified by the choice

$$\omega = \alpha = 1\,, \quad \beta = 0\,, \quad T(p) = -\frac{1}{m}\,\delta_+(p) \tag{40}$$

in (23), with $\delta_+(p) = \delta_+^0(p)$. The corresponding T^+ propagators are

$$\left.\begin{aligned} \tau^+_{\mu\nu}(p) &= \Big(-g_{\mu\nu} + \tfrac{p_\mu p_\nu}{p^2+i\epsilon}\Big)\,\tfrac{i}{2\pi(p^2-m^2+i\epsilon)} \\ \tau^+_{II}(p) &= -\tfrac{i}{2\pi(p^2+i\epsilon)} \\ \tau^+_{\mu I}(p) &= -\tau^+_{I\mu}(p) = \tfrac{p_\mu}{2\pi\, m\,(p^2+i\epsilon)} \\ \tau^+_{RR}(p) &= \tfrac{i}{2\pi(p^2-M^2+i\epsilon)} \end{aligned}\right\}\,. \tag{41}$$

The τ^- are obtained from τ^+ by the replacements $(i \to -i,\ p \to -p)$.

Notice that now the propagator $\tau^+_{\mu\nu}$ has a nice, renormalizable, large-p behavior, at the price of introducing the ghost factor$(p^2)^{-1}$. Unfortunately, that does not mean that the theory has become renormalizable. The bad UV behavior has merely been shifted to the mixed A-I propagators. But here the desired cancellations are easier to prove than in the U-gauge.

Before attacking this problem we must decide how the physical content of the model presents itself in the new gauge. Remember that in the U-gauge the physical state space is generated from the vacuum by applying polynomials in the gauge invariant fields $F_{\alpha\beta}$ and Ψ. Since the physical content of the theory must be gauge invariant, the same will be true in the R-gauge: the physical state space $\mathcal{V}_{ph}$, which is now a proper subspace of the full state space, is generated from the vacuum by applying polynomials in $F_{\alpha\beta}$ and Ψ. $\mathcal{V}_{ph}$ can again be reconstructed by the Wightman reconstruction theorem from the W-functions of these physical fields only. Only in these 'physical' W-functions are we really interested, hence only for them need we prove renormalizability.

The graph representation of the physical fields is clear. To obtain an external $F_{\alpha\beta}$ propagator, simply replace the factor $(-g_{\mu\nu}+p_\mu p_\nu/p^2)$ of an external A_μ line by $i\,(p_\alpha g_{\beta\nu}-p_\beta g_{\alpha\nu})$, the index ν belonging to the adjacent internal vertex. A $\Psi(p)$ factor is represented as a sum of three terms, an ordinary external R-line plus two external 2-prong vertices representing the composite fields R^2 and I^2 in (14). Both these composite vertices carry the vertex factor $g/(\sqrt{2\pi}^3 m)$.[g]

We turn now to the promised proof of the cancellation of UV dangerous terms. The basic idea is the following. Consider a I–A_μ cross propagator

$$w_{I\mu}(p) = i\,\frac{p_\mu}{m}\,\delta_+(p)\ ,$$

derived by (22) from the free 2-point function $\langle I(p)\,A_\mu(q)\rangle_0$. The end vertex of the corresponding cross line corresponds to a term in $\mathcal{R}^\mu(q)$. Summing over all these terms we obtain (with $q=-p$)

$$\delta_+(p)\,\Big(-i\,\frac{q_\mu}{m}\,\mathcal{R}^\mu(q)\Big) = -\delta_+(p)\,\mathcal{R}_I(q) \tag{42}$$

by the consistency condition (18). This means that we can replace the UV nice vertex sum $\mathcal{R}^\mu$ by the equally UV nice $\mathcal{R}_I$, and the UV bad propagator $w_{I\mu}(p)$ by the UV nice $-w_{II}(p)$! Unfortunately, in this crude form the argument is incorrect. The $\mathcal{R}^\mu(q)$ vertex in question belongs to, let us say, a T^+ sector, which represents a time ordered function of its external vertices, including the $\mathcal{R}$ vertex with $\mathcal{R}^\mu$ considered a composite external field. But in x-space the propagator factor $-i\,q_\mu$ represents a derivation ∂_μ acting not only on $\mathcal{R}^\mu(x)$ but also on the step functions occurring in the definition of the T-product. Hence we must expect that the relevant quantity

$$\frac{-i\,q_\mu}{m}\,\tau^+\big(\mathcal{R}^\mu(q)\cdots\big) + \tau^+\big(\mathcal{R}_I(q)\cdots\big) = \frac{1}{m}\,\tau^+\big(\mathcal{F}(q)\cdots\big)$$

does not vanish but is given as a sum of contact terms.[h] Luckily it turns out that these contact terms are not present if the sector in question contains

[g]The Fourier transform of the field $\varphi(x)$ is defined as $\varphi(p) = (2\pi)^{-5/2}\int dx\, e^{ipx}\varphi(x)$.
[h]The explicit form of this relation is known as a Ward-Takahashi identity.

only gauge invariant external fields. This is established by an explicit study of the graphs in question. Consider first the case that the $\mathcal{R}(q)$ vertices are those coming from L_3. Then the $\mathcal{F}_3$ occurring on the right-hand side is the Fourier transform of the expression (19). Consider a R-line with momentum k issuing from the vertex in question. Its denominator $(k^2 - M^2)^{-1}$ is cancelled by the numerator $(k^2 - M^2)$ coming from the first term in (19). Thus this first term leads to an amputation of the adjoining R-line, and a corresponding fusion of its two end vertices (internal or external) into a single vertex with more lines. The second term in (19) produces the same effect on I- and A-lines starting from the $\mathcal{F}$ vertex. In this way we obtain a considerable number of fused vertices, among which extensive cancellations occur. And the remaining fused vertices cancel against the L_4 terms in $\mathcal{F}(q)$. The actual verification of these cancellations is completely elementary but rather lengthy and tedious on account of the large number of different vertices to be considered (see (12), (13), (20)). The remarkable thing is, however, that these cancellations happen locally in the graphs in the immediate neighborhood of the q-end of the cross line in question, involving only that end vertex and its nearest neighbors, no matter how large the full sector may be. As a result we can, as proposed, drop our bad I–A_μ cross line and replace it by the negative of a good I–I line. The same argument, now used for the starting point, applies of course to a A_μ–I cross propagator. It may also be replaced by the negative of a I–I propagator. By this we end up with two negative I–I propagators for a given position of an appropriate line, plus the positive I–I propagator present from the beginning. The net effect is that we drop the dangerous mixed cross propagators and change the sign of the I–I cross propagators without changing our physical W-functions.

In this consideration we have assumed that the internal propagators in the sectors involved still have the original R-gauge form, and that the same applies to other cross propagators possibly involved in the cancellations. But the remarkable and lucky fact is that the said cancellations also occur if we have already effected the changes of rules explained above inside the sectors in question and in some of the cross lines, i.e. if we have already dropped there the mixed propagators and changed the signs of the I–I propagators. This enables us to prove the following

Theorem. *If in the graph rules of the R-gauge we omit the mixed A_μ–I lines and change the signs of the I–I propagators, then the resulting Wightman functions and related (partially or fully time ordered) functions of the physical fields $F_{\alpha\beta}$, Ψ, remain unchanged.*

Notice that the new graph rules arrived at in this way are those of the case $\alpha = 0$ ('Landau gauge') of the conventional L_α approach, thus confirming the perturbative validity of that approach. These new graph rules are clearly renormalizable in the sense of power counting. In fact, they are also renormalizable in the stricter sense that the necessary subtractions can be fully absorbed into renormalizations of the masses m, M, the coupling constant g, and the field normalizations. But the proof of this is quite involved and lies outside the scope of the present work.

The proof of the Theorem is inductive with respect to the order σ of perturbation theory. It consists of the following points.

(1) *The theorem is correct for* $\sigma \leq 2$. This is easily established by explicit calculation.

(2) *If the theorem is true for the 2-sector functions* $\big(\Omega, T^-(\cdots)\, T^+(\cdots)\, \Omega\big)_\sigma$, *then it is true for all n-sector functions* $\big(\Omega, T_1^\pm(\cdots) \cdots T_n^\pm(\cdots)\, \Omega\big)_\sigma$, *in particular the W-functions, with the same fields.* This is so because all these functions are in x-space boundary values of the same analytic function.[i] The reason for this is that, first, all permuted W-functions of a given set of fields are boundary values of a single analytic function (see,[7] Theorem 3-6), and that, second, any n-sector function is locally equal to a permuted W-function, wherever all x_i^0 are different and, because of Lorentz invariance, even where all x_i are different.

(3) Amputate the considered functions by multiplying them with $(p^2 - m^2)$ for factors $F_{\alpha\beta}(p)$, $(p^2 - M^2)$ for factors $\Psi(p)$. *Then the theorem is true for the full functions if it is true for the amputated ones.* This is so because we know precisely how to reconstruct the full functions from the amputated ones.

(4) *The theorem is true for the amputated 2-sector functions of order* σ. This is seen by noticing that in the corresponding 2-sector graphs both sectors are of orders ϱ with $0 < \varrho < \sigma$, so that the inductive hypothesis is applicable to them: the new rules can be used inside these sectors. Then the cross propagators linking them can also be changed to the new form by the arguments related above.

---

[i]Strictly speaking this is not true at points where two arguments in the same $T^\pm$ factor coincide. But this is of little concern because it does not happen in the W-functions, which are the functions of central interest.

6. The Non-Abelian Case

The methods used for the abelian HKM can be extended to the non-abelian case. In this last section we will briefly describe, without details, this extension and its results in the case of the gauge group SU_2.

The fields of the model are a complex 2-vector $\Phi(x)$ with the scalar components $\phi_1(x)$, $\phi_2(x)$, and a triplet $A_1^\mu(x), \cdots, A_3^\mu(x)$, of real vector fields. The Lagrangian is[j]

$$L = -\frac{1}{4} F_{a,\mu\nu} F_a^{\mu\nu} + \left[\partial_\mu - g A_{b,\mu} T_b) \Phi\right]^* \left[(\partial^\mu - g A_c^\mu T_c) \Phi\right] + \mu^2 \Phi^*\Phi - \lambda (\Phi^*\Phi)^2 . \tag{43}$$

Here

$$F_a^{\mu\nu} = \partial^\mu A_a^\nu - \partial^\nu A_a^\mu - g \varepsilon_{abc} A_b^\mu A_c^\nu , \tag{44}$$

and

$$T_a = -\frac{i}{2} \sigma_a , \tag{45}$$

σ_a the Pauli matrices.

L is invariant under the infinitesimal gauge transformations

$$\left.\begin{array}{l} \Phi(x) \Rightarrow (1 + g \vartheta_a(x) T_a) \Phi(x) \\ A_a^\mu(x) \Rightarrow A_a^\mu + g \varepsilon_{abc} \vartheta_b(x) A_c^\mu(x) + \partial^\mu \vartheta_a(x) \end{array}\right\} \tag{46}$$

for infinitesimal real functions ϑ_a. In contrast to the abelian case, the field strengths $F_a^{\mu\nu}$ are not gauge invariant.

The corresponding field equations possess the 'vacuum solution'

$$\Phi = \frac{1}{\sqrt{2}} \begin{pmatrix} v \\ 0 \end{pmatrix} , \qquad A_a^{\mu\nu} = 0 \ \forall a , \qquad v = \frac{\mu}{\sqrt{\lambda}} , \tag{47}$$

which takes over the role of the abelian solution (4). The ϕ_i are replaced as fundamental fields by the real scalar fields $R(x)$, $I_a(x)$, defined by the ansatz

$$\Phi = \frac{1}{\sqrt{2}} \begin{pmatrix} v + R + i I_3 \\ -I_2 + i I_1 \end{pmatrix} . \tag{48}$$

And, as in the abelian case, we replace the coupling constants μ, λ, as parameters of the theory by

$$m = \frac{v g}{2} , \qquad M = \sqrt{2} \mu , \tag{49}$$

---

[j]We use the summation convention both for Minkowski indices $\mu, \ldots$, and group indices $a, \ldots$.

which turn out to be the (unrenormalized) masses of the gauge bosons and the Higgs particle respectively. The field equations of the model look exactly like (15)–(17), except that there are now three A_a^μ-equations and three I_a-equations, one for each value of the group index a. Correspondingly we get now three consistence conditions:

$$\mathcal{F}_a := \partial_\mu \mathcal{R}_a^\mu + m\,\mathcal{R}_{Ia} = 0 \ . \tag{50}$$

Wightman gauges can be defined and constructed like in the abelian case. We are here not concerned with maximal generality, but need only consider the U- and the R-gauge. The *U-gauge* can again be characterized by the gauge condition $I_a = 0$ for all a. Its surviving cross propagators are taken over from (33) as

$$w_{ab}^{\mu\nu}(p) = -\,\delta_{ab}\,(g^{\mu\nu} - m^{-2}p^\mu p^\nu)\,\delta_+^m(p)\ , \quad w_{RR}(p) = \delta_+^M(p)\ , \tag{51}$$

and similarly for the sector propagators. The *R-gauge* is again defined by the propagators (41), where the first three lines hold for a-a propagators for any value of the group index a, while the mixed a-b propagators with $a \neq b$ vanish.

The physical space $\mathcal{V}_{ph}$ is again equated with the state space $\mathcal{V}_U$ of the U-gauge. In order to turn this into a gauge invariant definition also usable in the R-gauge, we must again produce $\mathcal{V}_U$ from the vacuum by applying gauge invariant fields. As one of these fields we use the Higgs field, which is now defined as

$$\Psi(x) = R(x) + \frac{g}{4m}\left[R^2(x) + I_a(x)\,I_a(x)\right] \ . \tag{52}$$

But the $F_a^{\alpha\beta}$ are no longer gauge invariant. However, we can replace them by gauge invariant fields, which we choose to be those introduced by Fröhlich et al.[2] As one of them we define

$$V_3^{\mu\nu}(x) := \frac{i\,g^2}{m^2}\,\Phi^*(x)\,T_a\,F_a^{\mu\nu}(x)\,\Phi(x)\ , \tag{53}$$

where Φ is expressed by (48) with $v = 2m/g$. In the U-gauge this becomes

$$V_3 = F_3 + \frac{g}{m}\,R\,F_3 + \frac{g^2}{4m^2}\,R^2\,F_3\ .$$

$V_2^{\mu\nu}$ is defined in the same way, except that the T_a are replaced by their cyclic permutation ($T_1 \to T_2,\ T_2 \to T_3,\ T_3 \to T_1$). Repeating this operation we obtain $V_1^{\mu\nu}$. By the same kind of arguments as used in Sect. 4 it can be shown that the restrictions to the U-gauge of these V_a, together with Ψ, indeed reproduce $\mathcal{V}_U$.

Hence again, the only W-functions of direct physical relevance are those containing only the physical fields Ψ, V_a, and only the renormalizability of these must be decided. And this is again easiest to achieve in the R-gauge. The method used is the same as in the abelian case. It turns out to be more complicated in its details. The main reason for this is that the simple form (19) of $\mathcal{F}_3$ is replaced by the more complicated expression

$$\begin{aligned}\mathcal{F}_{a3} &= g\,\varepsilon_{abc}\,A_{c\nu}\left[(\Box + m^2)\,A_b^\nu - \partial^\nu\partial^\mu A_{b\mu} + m\,\partial^\nu I_b\right]\\ &\quad+\frac{g}{2}\,I_a\,(\Box + M^2)\,R - \frac{g}{2}\,R\,(\Box I_a + m\,\partial_\mu A_a^\mu)\\ &\quad+\frac{g}{2}\,\varepsilon_{abc}\,I_b\,(\Box I_c + m\,\partial_\mu A_c^\mu)\ . \end{aligned} \tag{54}$$

Including this as a sum of composite external vertices in a sector in which the mixed A–I propagators are already eliminated, we find that

$$\partial_\mu A_a^\mu = 0\ ,$$

so that the corresponding terms in (54) can be dropped. But even so the terms in the first line of (54) do not have the desired fusing effect on the adjacent propagators. The factor $(p^2 - m^2)$ of the first term applied to an A^ν-A^λ propagator produces the ghost term $p^\nu p^\lambda/(m^2\,p^2)$, and the $p^\nu I_b$ term applied to a I-I propagator clearly does not remove its singularity at $p^2 = 0$. Hence, even if the fusing contributions do cancel like in the abelian case, there remains a non-fusing contribution. But the two offensive terms combine in such a way that they produce a ghost line ending in a new $\mathcal{F}$ vertex, now inside the sector, which fact allows using an inductive procedure leading to a simple result. It turns out that the undesirable non-fusing terms can be removed by the introduction of Faddeev-Popov ghost loops (FP loops).[1] Such a loop is a directed closed loop. Each line carries a propagator

$$\frac{i}{2\,\pi\,(p^2 + i\epsilon)}$$

(in a T^+ sector) and a group index $a, \cdots$. The loop contains only 3-line vertices with an A_c^ν line joining the loop. The vertex factor is

$$(2\sqrt{2\pi})^{-1}\,g\,\varepsilon_{abc}\,(p^\nu + q^\nu)$$

with p the loop momentum leaving the vertex, q that entering the vertex, and a and b are the indices of the lines respectively leaving and entering the vertex. And each such ghost loop contributes an extra factor -1.

We might then *conjecture* the following generalization of the Theorem of Sect. 5 to hold:

Change the graph rules of the R-gauge by omitting the mixed A_a^μ*-*I_a *propagators and changing the signs of the* I_a*-*I_a *propagators, and by admitting an arbitrary number of FP-loops. This procedure does not change the W-functions and related functions of the physical fields* $V_a^{\mu\nu}$, Ψ.

These conjectured rules are again the rules of the standard formalism in the Landau gauge.

The conjecture would be correct, if the fusing terms of $\mathcal{F}_3$ did lead to graph-local cancellations in analogy to the abelian case. This turns out to be the case for purely internal cancellations, that is if the end points of the fused lines are internal L_{int} vertices. But it is not true in all cases where external vertices (composite fields contributing to V_a) are involved. Therefore the equality of the physical W-functions in the Landau gauge and the R-gauge, and hence in the HKM in general, cannot be proved. This should not be interpreted as a weakness of our method. There are strong indications that the Landau gauge is indeed not physically equivalent to the HKM, if 'physical equivalence' is defined in our sense, not simply as the equality of the S-matrices.

As a result, there exists as yet no convincing proof of the full renormalizability of the non-abelian HKM.

References

1. L. D. Faddev, and V. N. Popov: *Phys. Letters 25B, 29 (1967).*
2. J. Fröhlich, G. Morchio, and F. Strocchi: *Nucl. Phys. B190, 553 (1981).*
3. C. Itzykson, and J.-B. Zuber. *Quantum Field Theory.* McGraw-Hill, New York, 1980.
4. A. Ostendorf: *Ann. Inst. H. Poincaré 40, 273 (1984).*
5. O. Steinmann: *Commun. Math. Phys. 152, 627 (1993).*
6. O. Steinmann: *Perturbative QED and Axiomatic Field Theory.* Springer, Berlin, 2000.
7. R. F. Streater, and A. S. Wightman: *PCT, Spin and Statistics, and All That.* Benjamin/Cummings, Reading MA, 1978.
8. S. Weinberg: *The Quantum Theory of Fields, Vol. 2.* Cambridge U. Press, Cambridge, 1996.
9. W. Zimmermann, in: *Lectures on Elementary Particles and Quantum Field Theory (ed. S. Deser et al.).* MIT Press, Cambridge MA, 1971.

LARGE REGULAR QCD COUPLING AT LOW ENERGY?

DMITRY V. SHIRKOV*
Bogoliubov Laboratory of Theoretical Physics, Joint Institute for Nuclear Research
141980, Dubna, Russia
**E-mail: shirkovd@theor.jinr.ru*

The issue is the expediency of the QCD notions' use in the low energy region down to the confinement scale, and, in particular, the efficacy of the QCD invariant coupling $\bar{\alpha}_s(Q^2)$ with a minimal analytic modification in this domain. To this goal, we overview a quite recent progress in application of the ghost-free Analytic Perturbative Theory approach (with no adjustable parameters) for QCD in the region below 1 GeV. Among them the Bethe-Salpeter analysis of the meson spectra and spin-dependent (polarization) Bjorken sum rule.

The impression is that there is a chance for theoretically consistent and numerically correlated description of hadronic events from the Z_0 till few hundred MeV scale by combination of analytic pQCD and some explicit non-perturbative contribution in the spirit of duality.

This is an invitation to practitioner community for a more courageous use of ghost-free models for data analysis in the low energy region.

Keywords: Quantum field theory; Quantum chromodynamics; Renormalization group

1. The pQCD Overview

QCD effective coupling $\bar{\alpha}_s$. Common perturbative QCD (pQCD) based upon Feynman diagrams starts with power expansion in $\alpha_s = g_s^2/4\pi \sim 0.1 - 0.4$, the strong interaction parameter analogous to the QED fine structure constant.

In QFT, an important physical notion is an invariant (or effective, or running) coupling function $\bar{\alpha}(Q)$, first mentioned in the QED context by Dirac (1933). In the current practice it was introduced in the basic renormalization group papers of mid-50s.[1]

The one-loop invariant QCD coupling sums up leading order (LO) logs into a geometric progression (with the Bethke[2] convention for the β_k coefficients)

$$\bar{\alpha}_s^{(1)}(Q) = \frac{\alpha_s(\mu)}{1+\alpha_s(\mu)\beta_0\,\ln(\frac{Q^2}{\mu^2})} = \frac{1}{\beta_0\,L}\,,\quad L=\ln\left(\frac{Q^2}{\Lambda^2}\right);\ \beta_0=\frac{33-2\,n_f}{12\pi}>0\,. \tag{1}$$

At the high enough energy (small distance), the QCD interaction diminishes $\bar{\alpha}_s(Q)\sim 1/\ln Q\to 0$ as $Q/\Lambda\to\infty;\quad r\,\Lambda\to 0$. This feature is the famous phenomenon of Asymptotic Freedom.

At the same time, eq.(1) obeys unphysical singularity (Landau pole) $\sim 1/(Q^2-\Lambda^2)$ in low-energy physical region at $|Q|=\Lambda\sim 400\,\text{MeV}$. Transition to the 2-loop case does not resolve the issue.

The asymptotic freedom behavior $1/\ln Q$ remains dominant in the 2-loop or Next-to-Leading-Order (NLO) case. Here, an explicit expression for $\bar{\alpha}_s$ obtained by iterative approximate solving[1,3] of differential RG equation, can be written down in a compact, the "denominator form" (as it was recently motivated in[4])

$$\bar{\alpha}_s^{(2)}(Q) = \frac{1}{\beta_0\,L+\frac{\beta_1}{\beta_0}\ln L}\,;\qquad \beta_1(n_f)=\frac{153-19n_f}{24\pi^2} \tag{2}$$

with values $\beta_0(4\pm1)=0.663\mp0.053\,;\,\beta_1(4\pm1)=0.325\mp0.085\,$.

The QCD scale in the $\overline{\text{MS}}$ scheme $\Lambda^{(n_f)}=\Lambda^{(n_f)}_{\overline{\text{MS}}}$, as obtained from the data happens to be close to the confinement scale $\Lambda^{(4\pm1)}\sim 300{\mp}100$ MeV $\simeq 2\,m_\pi$ or $R_\Lambda\sim 10^{-13}\,$cm.

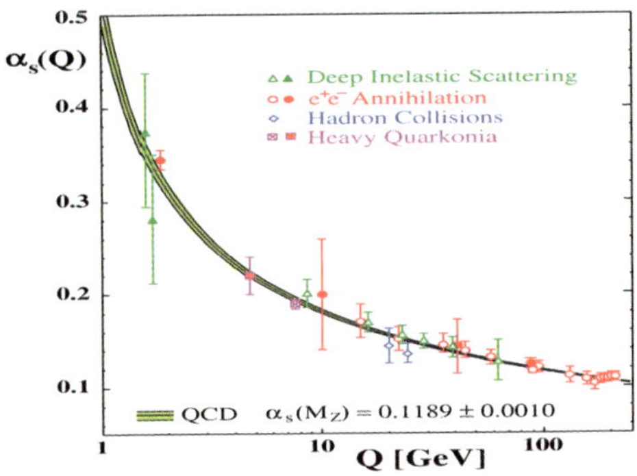

Fig. 1. *Effective QCD coupling correlating all the data in the range from few GeV up to few hundred GeV. The solid curve correspond to 2-loop, NLO case. Taken from the Bethke paper.*[2]

According to Bethke,[2] the 2-loop pQCD approximation (2) turns out to be sufficient for numerical correlation of several dozen of various experiments. Indeed, Fig.1 gives the evidence for *the two-loop pQCD triumph*: the NLO theoretical curve describes quite accurate - within the current experimental and theoretical errors – all the data in the energy range from 5 up to a few hundreds GeV.

However, below 5 GeV the correlation is not so persuasive. Moreover, in this region the data on Fig.1 (as well as in the corresponding PDG[5] plot) are rather scanty. The reason is the well known "Landau pole trouble".

As it is well known, widely used expressions for effective QCD coupling (like eqs.(1),(2); see also eq.(7) in Ref.[2]) and eq.(9.5) in Ref.[5]) suffer from spurious singularities, like Landau pole, in the LE physical region at $|Q| \sim \Lambda^{(3)} \sim 400\,\mathrm{MeV}$. This trouble embarrasses the data analysis by pQCD theory below a few GeV.

Unphysical pQCD singularity vs. lattice data. Meanwhile, lattice simulation results[6–8] testify the regularity of $\bar{\alpha}_s(Q)$ behavior in the region below 1 GeV. Indeed, as it was summarized in papers,[9,10] all the lattice data indicate smooth growth of α_s till specific scale $Q = Q_* \sim 400 - 500\,\mathrm{MeV}$ (that is close to $\Lambda^{(3)}$) with typical values $\bar{\alpha}_s(Q_*) \sim 0.5 - 0.8 < 1$.

This means that commonly used iterative solutions of RG eqs., like (1), (2) not only can but should be modified to correlate with lattice data.

Modifications of "Common pQCD" in LE domain. Several attempts to elude the pQCD singularities have been undertaken since 80s. Among them the straightforward freezing,[11] and some others more sophisticated, like glueball mechanism[12] and exponential modification.[13] All of them introduce some model parameters.

Meanwhile, in the mid-90s, an elegant way (free of additional parameters) to resolve this issue was proposed by Igor Solovtsov and collaborators[14–16] on the basis of the causality principle implemented in the form of the Källen – Lehmann analyticity for the QCD coupling $\bar{\alpha}_s(Q^2)$. Then, on the ground of Q^2-analyticity, consistent scheme known as Analytic Perturbation Theory (=APT) has been devised[17–19] during last decade.

Below, we give resume of the APT essence (Sect.2) and its application to data (Sect.3) in the above-mentioned troublesome region. These results produce a hope that the Bethke's *issue of two-loop α_s adequacy* can be proliferated to one more order of magnitude – down to few hundreds MeV with help of analytically modified QCD coupling and some additional non-perturbative means in the spirit of duality.

2. Analytic Perturbation Theory

Here, we start with short a sketch of APT. For detail see the review papers.[20–24]

2.1. *APT - General*

As it is well known, the 1st step of improving straightforward renormalized PT result is supplied by the RG Method[1] which allows one to restore the correct structure of singularity of a partial solution; in the QFT case – the correct UV and IR asymptotics. Its essence is a technique of reconstracting the so called *renormalization*[a] *invariance.* In QFT, the RG-improved results obey a drawback, the unphysical singularity.

In the latter case, the 2nd step, a further improving of RG-invariant PT solution should be used. Its main idea, imposing the *analyticity imperative* – that in turn stems out of the causality condition – was first formulated in the QED context.[27] A more elaborated QCD counterpart, the APT algorithm, is based on the following principles :

- Causality, that results in the analyticity of the effective coupling in the complex Q^2 plane a là Källen-Lehmann representation[b]

$$\bar{\alpha}_s(Q^2) \to \alpha_{\rm E}(Q^2) = \frac{1}{\pi}\int_0^\infty d\sigma\,\frac{\rho(\sigma)}{\sigma + Q^2 - i\epsilon}\,.$$

  This property provides the absence of spurious singularities.
- Correspondence with perturbative RG-improved input by proper defining $\rho(\sigma) = \mathrm{Im}\,\bar{\alpha}_s(-\sigma)$.
- Representation invariance, i.e., compatibility with linear integral transformations, like transition from the Euclidean, transfer momentum, picture to the Minkowskian, c.o.m. energy, one:

$$\alpha_E(Q^2) = Q^2\int_0^\infty \frac{\alpha_M(s)\,ds}{(s+Q^2)^2}$$

  (or the Fourier transition from $\alpha_E(Q)$ to its Distance image $\alpha_{\rm D}(r)$) that yields[28] non-power functional expansions for observables – see, below eqs.(4),(5).

---

[a]Or, more exactly, by the *reparameterization invariance*[25] of a partial solution. Recently, this RG technique has been devised for a class of boundary value problems of classical mathematical physics.[26]

[b]For some cases it is implemented in a form of the Jost-Lehmann (see Sec.4 in Ref.[20]) representation.

2.2. *The APT Algorithm*

Euclidean functions. Euclidean ghost-free expansion functions[28] are defined

$$\mathcal{A}_n(Q^2) = \int_0^\infty \frac{\rho_n(\sigma)}{\sigma + Q^2}\, d\sigma, \quad \rho_n(\sigma) = \frac{1}{\pi}\, \mathrm{Im}[\bar{\alpha}_s(-\sigma - i\varepsilon)]^n \tag{3}$$

via powers of $\bar{\alpha}_s$. They form non-power set of functions $\{\mathcal{A}_k(Q^2)\}$ that serves as a basis for modified non-power PT expansion of RG invariant objects in the Q picture, like Adler D-function. The first of these functions can be treated as an Euclidean APT coupling $\alpha_{\mathrm{E}}(Q^2) = \mathcal{A}_1(Q^2)$. In the one-loop case

$$\alpha_{\mathrm{E}}^{(1)}(Q^2) = \frac{1}{\beta_0}\left[\frac{1}{\ln(Q^2/\Lambda^2)} + \frac{\Lambda^2}{\Lambda^2 - Q^2}\right]$$

it differs from usual one $\alpha_s(Q^2)$ by the term $\sim 1/(\Lambda^2 - Q^2)$ that subtracts the singularity.

Here, higher expansion functions are related by elegant recurrent relation

$$\mathcal{A}_{n+1}^{(1)}(Q^2) = -\frac{1}{n\beta_0}\frac{d\,\mathcal{A}_n^{(1)}(Q^2)}{d\,\ln Q^2}\,.$$

Minkowskian expansion functions are connected[14,17,28] with the Euclidean ones by contour integral and the reverse "Adler transformation"

$$\mathfrak{A}_k(s) = \frac{i}{2\pi}\int_{s-i\varepsilon}^{s+i\varepsilon} \frac{dz}{z}\,\mathcal{A}_k(-z)\,; \quad \mathcal{A}_k(Q^2) = Q^2 \int_0^\infty \frac{\mathfrak{A}_k(s)\,ds}{(s+Q^2)^2}\,.$$

Minkowskian APT coupling $\alpha_M(s) = \mathfrak{A}_1(s)$ in the 1-loop case

$$\alpha_M^{(1)}(s) = \frac{1}{\pi\beta_0}\arccos\frac{L}{\sqrt{L^2+\pi^2}}\Bigg|_{L>0} = \frac{1}{\pi\,\beta_0}\arctan\frac{\pi}{L}\,, \quad L = \ln(s/\Lambda^2)$$

quantitatively is close to Euclidean APT one; see Fig.2.

Non-power APT - Loop and RS Stability. In APT, instead of universal power-in-$\bar{\alpha}_s$ ($\bar{\alpha}_s(Q^2)$ or $\bar{\alpha}_s(s)$) series

$$d_{pt}(Q^2/s) = d_1\bar{\alpha}_s(Q^2/s) + d_2\,\bar{\alpha}_s^2 + 0(\bar{\alpha}_s^3)\,,$$

one should use for each representation its own particular non-power expansion

$$d_{\mathrm{an}}(Q^2) = d_1\,\alpha_E(Q^2) + d_2\,\mathcal{A}_2(Q^2) + d_3\,\mathcal{A}_3(Q^2) + \dots\,, \tag{4}$$

$$r_\pi(s) = d_1\,\alpha_M(s) + d_2\,\mathfrak{A}_2(s) + d_3\,\mathfrak{A}_3(s) + \dots \tag{5}$$

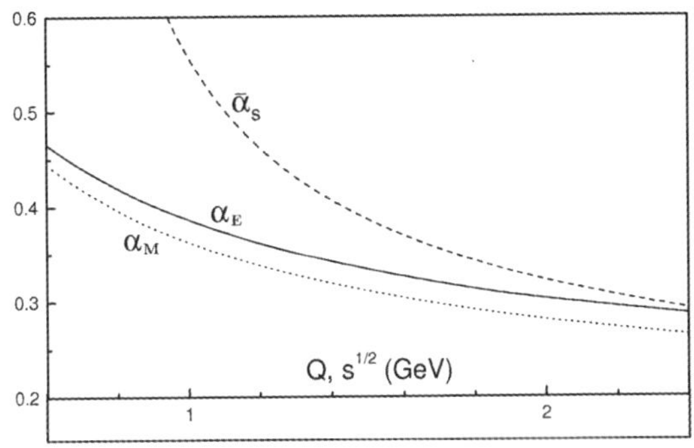

Fig. 2. *Comparison of singular $\bar{\alpha}_s$ coupling with Euclidean α_E and Minkowskian α_M in a few GeV region. Taken from paper.*[18]

that provides better loop convergence and practical RS independence of observables.

The 3rd terms in (3), (4) contribute into observables less than 5%.[21] Again the 2-loop (NLO) level is practically sufficient.

Fractional APT. In the computation of higher-order corrections to inclusive and exclusive processes one deals with non-integer fractional powers of QCD coupling. For such a case, special *fractional* generalization has been devised[29] and successively applied to pion form factor[30] and to the Higgs boson decay into a $b\bar{b}$ pair.[31]

2.3. *APT functions at LE region*

Comparison of APT Euclidean α_E and Minkowskian α_M couplings reveals that below 2-3 GeV scale they, being close to each other, differ seriously from the common singular $\bar{\alpha}_s$ – see Fig.2. Qualitatively, the same is true for higher expansion functions.

The APT RenormScheme- and loop- stability. In Fig.3, we give Euclidean APT coupling in the one-, two- and three-loop (NNLO) approximations taken in the $\overline{\text{MS}}$ scheme.

A beautiful feature of these curves is their relative loop stability. The two-loop curve below 1 GeV differs from the three-loop one by less than 3 per cent. Hence, the APT two-loop (NLO) curve is accurate enough for practical use at three-flavor region. This correlates with the above mentioned Bethke's conclusion for the higher energies.

In a real QCD case, one has to take into account the proper conjunction of regions with different values of effective flavour number n_f. This, a bit

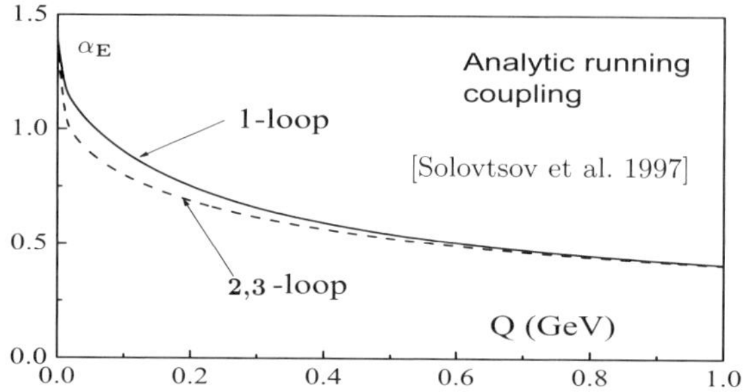

Fig. 3. *Euclidean APT coupling α_E in the 1-, 2- and 3-loop cases for the $\overline{\mathrm{MS}}$ scheme. Taken from paper.*[18]

subtle, issue was elaborated in paper.[19] From the practical point of view, one should use common matching conditions for recalculation of adjacent $\Lambda^{(n_f)}$ values for the quark threshold crossing. Resulting Euclidean functions $\mathcal{A}_k$ turn out to be smooth in the threshold vicinity, while Minkowskian ones $\mathfrak{A}_k$ remaining continuous have jumps in derivatives.

Recall here, that all this is valid for simple APT functions *without additional parameters*. Such a version is known as a *minimal APT*. Below, we shortly mention its massive generalization which contain an additional fitting parameter.

The "massive" APT modification. A quite natural ansatz has been added to minimal APT formalism in paper.[32] There, the lower limit in the Kállen-Lehmann integral Eq.(3) was changed from zero to $m^2 > 0$. This parameter, reminiscent of pion mass m_π squared is an additional one that can be used for the data fitting – see Fig.4.

3. Low Energy APT Application

The APT approach during the decade of it existence has been applied to the number of low energy (above 1-2 GeV) hadronic observables. One has to mention here sum rules,[33,34] $e^+ e^-$ inclusive hadron annihilation,[35] τ[36] and Υ decays,[37] above mentioned formfactors,[30,31] and some others. For detail one could address to review papers.[21,22] Below, we shortly overview quite fresh APT applications to processes in a rather low energy region $\lesssim 1\,\mathrm{GeV}$.

3.1. *APT and bosonic spectrum*

APT + Bethe-Salpeter formalism. Here, we present briefly a summary of recent analysis[38] of the meson spectrum by combination of the Bethe-Salpeter equation for the $(q, \bar{q})$ system with the APT approach.

By use of the 3-dimensional reduction, the BS eq. takes the form of an eigenvalue equation for a squared bound state mass

$$M^2 = M_0^2 + U_{\rm OGE} + U_{\rm Conf}\,,$$

with $M_0 = \sqrt{m_1^2 + \mathbf{k}^2} + \sqrt{m_2^2 + \mathbf{k}^2}$ – kinematic term, $U_{\rm Conf}$ – *confining potential*, $U_{\rm OGE}$ – one-gluon exchange potential $\sim$ QCD coupling

$$\langle \mathbf{k}|U_{\rm OGE}|\mathbf{k}'\rangle = \alpha_s(\mathbf{Q}^2)\, M_{\rm OGE}(\mathbf{Q} = \mathbf{k} - \mathbf{k}'\,, \mathbf{k})\,.$$

For a given bound state a, one has (for details see Refs.[40])

$$m_a^2 = \langle \phi_a|M_0^2|\phi_a\rangle + \langle \phi_a|U_{\rm OGE}|\phi_a\rangle + \langle \phi_a|U_{\rm Conf}|\phi_a\rangle\,.$$

These two relations allow one to extract $\alpha_s(Q_a^2)$ values for a low enough momentum transfer region $100\,\text{MeV} < Q_a < 1\,\text{GeV}$.

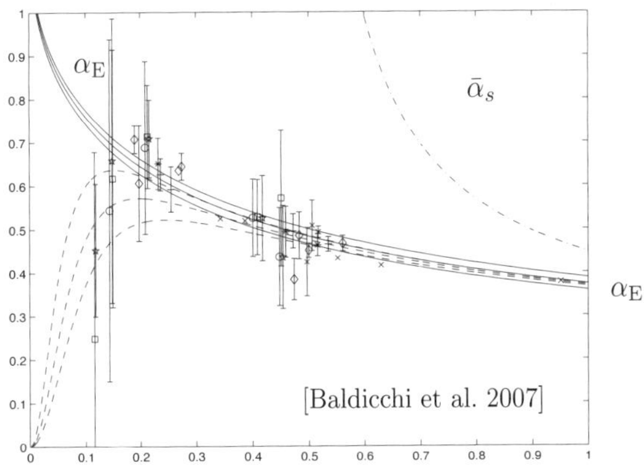

Fig. 4. *Comparison of α_s from meson spectrum (points with error bars) and 3-loop α_E at $\Lambda^{(3)}_{n_f=3} = (417\pm42)$ MeV (3 solid curves). Singular 3-loop $\bar{\alpha}_s$ coupling (dot-dashed) is excluded by data. Dashed lines correspond to the massive APT version[32] with $m \sim 40\,\text{MeV}$. Taken from the paper.[38]*

Results of α_s extraction from bosonic spectrum are given in Fig.4. One sees that meson spectrum data roughly follow a bunch of three α_E curves for $\Lambda^{(3)}_{n_f=3} = (417\pm42)\,\text{MeV}$ corresponding to the 2006 world average $\bar{\alpha}_s(M_Z^2) = 0.1189 \pm .0010$. There is also a slight hint on the tendency for

BS-extracted $\alpha_s$values at $Q < 200\,\mathrm{MeV}$ to diminish in the IR limit. The dashed curves on Fig.4 just correspond to this possibility. However, in our opinion, this scenario is supported only by data from the D and F orbital excitations of the $(q, \bar{q})$ system. They have big error bars and some of them are subject to a doubtful interpretation. If we exclude higher states and limit ourselves to the S and P ones, the resulting picture will change.

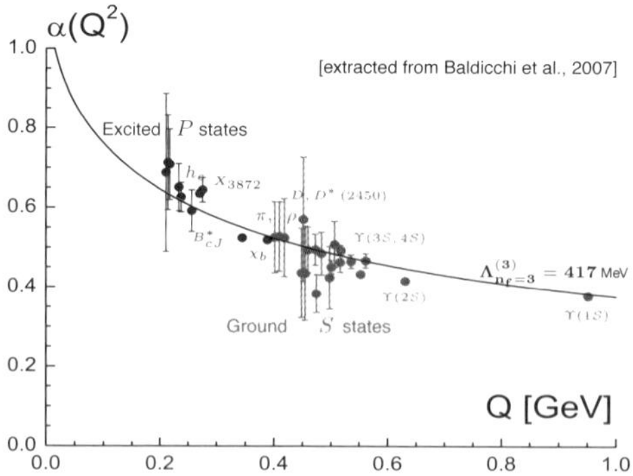

Fig. 5. The APT $\alpha_{\mathrm{E}}(Q^2)$ coupling correlated with the world average vs. α_s^{exp} from the S,P states of the $(q, \bar{q})$ system. Evidence for evolution below 500 MeV.

APT vs S and P data. In the last Fig.5, we show the picture without higher orbital D and F excitations. This limited set of data with small error bars quite nicely follows just the APT coupling curve with the world average $\Lambda^{(3)}_{n_f=3} = 417$ MeV value.

3.2. *Bjorken sum rule*

The analysis of recent 2006 Jefferson Lab data on the Bjorken Sum Rule for the moments of spin-dependent structure function Γ_{p-n} at $0.1 < Q^2 < 3\,\mathrm{GeV}^2$ in the NNLO approximation was recently performed.[41] Higher twists (HT) values extracted within the APT provide evidence for better convergence of HT series as compared to the standard pQCD. As a final result, a reasonable quantitative description of the data down to 350 MeV was achieved.

Together with the meson spectra evidence Fig.5, this produces an impression that minimal APT allows one to enlarge the domain of analytic perturbative QCD (supported by a transparent non-perturbative elements) description of hadronic events down to few hundreds MeV.

4. APT in QCD: Conclusion

Meson spectrum data analyzed by the Milano BS-technique with the one-gluon exchange potential and confinement ansatz result[39] in extraction the QCD coupling $\bar{\alpha}_s(Q)$ values in the LE domain of momentum transfer $Q < 1\,\text{GeV}$. In a recent research it was shown[38,40] that the use of ghost-free analytic QCD Euclidean coupling α_E in this analysis yields rather an intriguing correlation (shown in Figs.4 and 5) of the "meson spectrum α_s values" in the region $250\,\text{MeV} \lesssim Q < 1\,\text{GeV}$ with the world average $\bar{\alpha}_s(M_Z^2)$.

Along with this, the arena for the APT non-power expansion results for the Bjorken sum rule[41] is also ranging down as far as to the $\sim 300-400\,\text{MeV}$ scale.

Both the results –

- exclude common α_s singular behavior and smooth "freezing" below 1 GeV,
- support minimal APT extension of pQCD, giving hope for a quasi-perturbative consistent quantitative picture from 200 GeV to 200-300 MeV.

Due to this, there appears a chance for the real possibility of consistent theoretical analysis of hadronic processes in the low-energy region, the chance that is based on two elements:
– the procedure of getting rid of spurious singularities, by some low-energy modification of pQGD, like the APT one;
– addition of some appropriate non-perturbative elements in the spirit of parton-hadron duality, like confinement ansatz and higher twist contribution.

We appeal to the QCD practicing community for a more regular use of ghost-free QCD coupling models for data analysis in the low energy region below 1 –2 GeV. Just in this region theoretical errors quite often exceed the experimental ones.

Acknowledgements

The author is grateful to Professor Wolfhart Zimmermann and Prof. E. Seiler for hospitality in MPI, Muenchen. The useful discussion with Drs. A.Bakulev, S. Bethke, S.Mikhailov, O.Solovtsova, N.Stefanis and O.Teryaev is sincerely acknowledged. This work was supported in part by RFBR grant 08-01-00686, the BRFBR (contract F08D-001) and RF Scientific School grant 1027.2008.2.

References

1. N. N. Bogoliubov and D. V. Shirkov, Doklady AN SSSR, **103** 203, 391 (1955).
   N. N. Bogoliubov and D. V. Shirkov, *Sov.Phys.JETP*, **3** (1956) 57.
   N. N. Bogoliubov and D. V. Shirkov, *Nuovo Cim.* **3** (1956) 845.
2. S. Bethke, *Prog.Part.Nucl.Phys.* **58** (2007) 351-386; hep-ex/0606035.
3. D. V. Shirkov, *Theor. Mat. Fiz.(USA)* **49** (1981) No.3, 1039-43.
4. D. V. Shirkov, Nucl.Phys.B(Proc.Suppl.) **162**:33-38 (2006); hep-ph/0611048.
5. W.-M. Yao *et al.*, Journ. Phys., **G 33**, 1 (2006).
6. L. Alkofer, L. von Smekal, *Phys. Repts.* **353** (2001) 281; hep-ph/0007355.
7. Ph. Boucaud *et al.*, *Nucl.Phys.Proc.Suppl.* **B106** (2002) 266; hep-ph/0110171.
   Ph. Boucaud *et al.*, *JHEP* 0201 (2002) 046; hep-ph/0107278.
8. J. I. Skullerud, A. Kizilersu, A. G. Williams, *Nucl.Phys.Proc.Suppl.* **B 106** (2002) 841; hep-lat/0109027.
   J. Skullerud, A. Kizilersu, JHEP 0209 (2002) 013; hep-ph/0205318.
   J. I. Skullerud, *et al.*, JHEP 0304 (2003) 047: hep-ph/0303176.
9. See Sect. 2 in D. V. Shirkov, *Theor. Math.Phys.* **132** (2002) 1309; hep-ph/0208082
10. G. M. Prosperi, M. Raciti, C. Simolo, Prog.Part.Nucl.Phys. 58 (2007) 387-438; hep-ph/0607209.
11. G. Grunberg, Phys.Lett.**B 95**:70,(1980), Erratum-ibid.B110:501,1982
    G. Grunberg, Phys.Rev.**D 29** :2315, 1984.
12. Yu. A. Simonov, Nucl.Phys. **B 324** :67,(1989).
    A. M. Badalian: Phys.Rev. **D 65** :016004, (2002); hep-ph/0104097
13. G. Cvetic, Cr. Valenzuela, Phys.Rev. **D 77**: 074021, (2008); hep-ph/0710.4530.
14. H. F. Jones and I. L. Solovtsov, *Phys. Let.* **B 349** (1995) 519; hep-ph/9501344
15. D. V. Shirkov and I. L. Solovtsov, *JINR Rap. Comm.* **2** [76], 5 (1996); hep-ph/9604363.
16. D. V. Shirkov and I. L. Solovtsov, *Phys.Rev. Lett.* **79**, 1209 (1997); hep-ph/9704333.
17. K. A. Milton and I. L. Solovtsov, *Phys.Rev.*D **55**, 5295 (1997), hep-ph/9611438
18. I. L. Solovtsov and D. V. Shirkov, *Phys.Lett.* **B 442**, 344 (1998); hep-ph/971251.
19. D. V. Shirkov, *Theor. Math.Phys.* **127** 409 (2001); hep-ph/0012283.
20. I. L. Solovtsov, D. V. Shirkov, Theor.Math.Phys. **120**: 1220, (1999); hep-ph/9909305.
21. D. V. Shirkov, *Eur.Phys.J.* C **22** (2001) 331; hep-ph/0107282.
22. I. L. Solovtsov and D. V. Shirkov, Theor.Math.Phys. **150** (2007) 132; hep-ph/0611229;
23. G. Prosperi, Prog.Part.Nucl.Phys. **58** 387-438 (2007);hep-ph/0607209.
24. G. Cvetic, C. Valenzuela, "Analytic QCD: A Short review", USM-TH-227, Apr 2008. 10pp. hep-ph/0804.0872
25. D. V. Shirkov, *Sov.Phys.Dokl.* **27** (1982) 107.
26. V. F. Kovalev, D. V. Shirkov, *J.Phys.A* **39** (2006) 806.
27. N. N. Bogoliubov, A. A. Logunov, and D.V. Shirkov, *JETP* **37** (1959) 805.
28. D. V. Shirkov, *TMP* **119** (1999) 438; hep-th/9810246;

Lett.Math.Phys. **48** (1999) 135.
29. A. P. Bakulev, A. I. Karanikas, N. G. Stefanis, Phys.Rev. D72 (2005) 074015; hep-ph/0504275
30. A. P. Bakulev, S. V. Mikhailov, N. G. Stefanis, Phys.Rev. D72 (2005) 074014; hep-ph/0506311
31. A. P. Bakulev, S. V. Mikhailov, N. G. Stefanis, Phys.Rev. D75 (2007) 056005; hep-ph/0607040
32. A. V. Nesterenko, J. Papavassiliou, Phys.Rev. **D** **71**:016009, (2005); hep-ph/0410406.
A. V. Nesterenko, J.Phys. **G** **32** 1025,(2006); hep-ph/0511215.
33. K. A. Milton, I. L. Solovtsov, and O. P. Solovtsova, Phys. Rev. D, 60, 016001 (1999).
34. K. A. Milton, I. L. Solovtsov, O. P. Solovtsova, Phys. Lett. B **439**, 421 (1998); hep-ph/9809510.
35. D. V. Shirkov, I. L. Solovtsov, Proc. Workshop on e^+e^- Collisions from ϕ to J/Ψ March 1999, Eds. G. V. Fedotovich and S. I. Redin, Budker Inst. Phys., Novosibirsk, 2000, pp. 122-124; hep-ph/9906495.
36. K. A. Milton, O. P. Solovtsova, *Phys.Rev.* **D** 57 (1998) 5402-5409; hep-ph/9710316.
37. D. V. Shirkov, A. V. Zayakin Phys.Atom.Nucl. **70** :775-783, (2007): hep-ph/0512325.
38. M. Baldicchi *et al.*, *Phys.Rev.Lett.* **99** 242001 (2007), hep-ph/0705.0329.
39. M. Baldicchi, G. M. Prosperi, *Phys.Rev.* D **66**:074008, (2002), hep-ph/0202172;
AIP Conf.Proc. **756**:152-161,2005. : hep-ph/0412359
40. M. Baldicchi *et al.*, Phys.Rev.D **77**:034013 (2008); hep-ph/0705.1695
41. R. Pasechnik, D. Shirkov, O. Teryaev, in preparation.

THE DIHEDRAL GROUP AS A FAMILY GROUP

JISUKE KUBO*

Institute for Theoretical Physics, School of Mathematics and Physics
Kanazawa University, Kanazawa, 1192, Japan
**E-mail: jik@hep.s.kanazawa-u.ac.jp*

After a brief introduction into finite groups, practical tools for dealing with the dihedral group are developed. Then a recently proposed flavor model with a family group based on the binary dihedral group Q_6 is introduced. The predictions of the model on the Cabibbo-Kobayashi-Maskawa parameters are analyzed to investigate their testability at feature B-factory experiments.

Keywords: Finite groups; Family symmetries; Super B-factories

1. Introduction

It is widely believed that the standard model (SM) of elementary particles should be extended. One of the reasons is the Yukawa sector, because the most of the free parameters of the SM are involved in this sector, and the SM does not provide with a principle how to fix its structure. A natural way to provide with a principle for the Yukawa sector is the introduction of a family symmetry. A family symmetry is not necessarily adequate to explain the observed hierarchy of the fermion masses. It can however relates the fermion masses and mixing parameters. Recently, there have been a growing number of interests in family symmetries. The most of the recent papers deal with the large neutrino mixing, because a large mixing may be associated with a family symmetry. Here we are interested in models in which the notion of family is extended to the Higgs sector and family symmetry is not hardly broken at low energy.[1] This type of models can make specific predictions that may be testable at future experiments. Here we particularly pay attention to a supersymmetric flavor model with a Q_6 family group,[2] and consider its testability[3] by B factories such as SuperKEKB[4] and Super Flavour Factory.[5]

Before we come to the specific model, we will briefly outline the basic notions of finite groups, and then focus on the dihedral group. As we will see, we develop practical tools for dealing with the dihedral group, which, to our knowledge, can not be found in literature.

2. Finite groups

The classification of finite groups has been completed by D. Gorenstein (1983), and M. Aschbacher and S. Smith (1995), much later than the case of the continues group. Abelian finite groups are well-known and frequently used in particle physics. But except for the permutation group S_3 (which is isomorphic to the dihedral group D_3), non-abelian finite groups are not very much applied in particle physics [a]. They are rather known to solid state physicists or crystalogists. See for instance "Quantum mechanics" by Landau and Lifshitz,[7] in which various examples of finite groups are treated. For the following discussions the text book by Wu-Ki Tung[8] may be very useful.

2.1. *Character table*

The number of the group elements of a finite group is called the order g of the group. The most important quantity in finding irreducible representations (irreps) of a finite group is the character $\chi(G) = \text{Tr } D(G)$ of $G \in \mathcal{G}$, where G is an arbitrary element of the finite group $\mathcal{G} = \{ G_1, \ldots, G_g \}$, and $D(G)$ is a matrix representation of G. The elements can be grouped into a certain number of classes C_k, which are defined as

$$G\ C_k\ G^{-1} = C_k \tag{1}$$

for an arbitrary group element G. The character $\chi(G)$ depends only on the irrep and class. Since the number of distinct classes n_c coincides with the number of inequivalent irreps n_r, the character χ is an $n_c \times n_c$ matrix ${\chi^\mu}_k$, which is called the character table for the group $\mathcal{G}$. (The irreps are labeled by μ.) The following theorems play the central roles to complete a character table: (i) $n_r = n_c$. (ii) $\sum_{\mu=1}^{n_r} (d_\mu)^2 = g$, where d_μ is the dimension of the irrep labeled by μ. (iii) $\sum_{k=1}^{n_c} h_k\ ({\chi^\mu}_k)^*\ {\chi^\nu}_k = g\delta_{\mu\nu}$, where h_k is the dimension of the class C_k. (vi) $h_i \sum_{\mu=1}^{n_r}\ ({\chi^\mu}_i)^* {\chi^\mu}_j = g\delta_{ij}$. (v) $h_i h_j {\chi^\mu}_i {\chi^\mu}_j = d_\mu \sum_{k=1}^{n_c}\ c_{ij}^k h_k\ {\chi^\mu}_k$, where $C_i C_j = \sum_{k=1^{n_c}}\ c_{ij}^k C_k$.

2.2. *Tensor products*

The decomposition of tensor products into irreps is essential in constricting an invariant Lagrangian. Let u_m^μ $(m = 1, \ldots, d_\mu)$ be the bases and transform

[a] One of the earliest papers in particle physics is.[6]

as $Gu^{\mu}_{m} = \sum_{n} D^{\mu}_{mn}(G)u^{\mu}_{n}$. The useful tool for the decomposition is the projection operator

$$P^{\mu}_{lm} = \frac{d_{\mu}}{g}\sum_{G} D^{\mu}_{ml}(G)^{*}\, G, \tag{2}$$

which picks up u^{μ}_{l} from $f = \sum_{\gamma,n} c^{\gamma}_{n}u^{\gamma}_{n}$, i.e., $P^{\mu}_{lm}\, f = c^{\mu}_{m}u^{\mu}_{l}$.

This projection operator can be used to decompose a tensor product into irreps. Let u^{μ} and v^{ν} be the bases transforming as $Gu^{\mu}_{m} = \sum_{m'} D^{\mu}_{mm'}(G)u^{\mu}_{m'}$ and $Gv^{\nu}_{n} = \sum_{n'} D^{\nu}_{nn'}(G)v^{\nu}_{n'}$, respectively. Then consider an arbitrary linear combination of bilinear products of u^{μ} and v^{ν}, i.e., $f = \sum_{m,n} k_{mn}u^{\mu}_{m}v^{\nu}_{n}$. Since $D^{\mu}\times D^{\nu}$ can be decomposed into a sum of D^{γ}'s, f can be expanded in terms of the bases w^{γ} corresponding to D^{γ}:

$$f = \sum_{m,n} k_{mn}u^{\mu}_{m}v^{\nu}_{n} = \sum_{\gamma,l} c^{\gamma}_{l}w^{\gamma}_{l}, \tag{3}$$

where w^{γ}_{l} can be written as a linear combination of the bilinear products $w^{\gamma}_{l} = \sum_{m,n} c^{\gamma}_{lmn}u^{\mu}_{m}v^{\nu}_{n}$. Then applying the projection operator on both sides of (3) we obtain

$$\begin{aligned} P^{\alpha}_{lp}\, f &= \sum_{m,n} k_{mn}\frac{d_{\alpha}}{g}\sum_{G} D^{\alpha}_{pl}(G)^{*}\, D^{\mu}_{mm'}(G)u^{\mu}_{m'}\, D^{\nu}_{nn'}(G)v^{\nu}_{n'} \\ &= c^{\alpha}_{p}w^{\alpha}_{l} = c^{\alpha}_{p}\sum_{m,n} c^{\alpha}_{lmn}u^{\mu}_{m}v^{\nu}_{n}. \end{aligned} \tag{4}$$

The r.h.s. of (4) can be explicitly calculated, so that there are $(d_{\alpha})^2$ equations which can be used to calculate the coefficients c^{α}_{lmn}. Similarly, one can construct a G invariant from $\sum_{l,m,n} k_{lmn}w^{\gamma}_{l}u^{\mu}_{m}v^{\nu}_{n}$ if we use $D(G) = \mathbf{1}$:

$$\sum_{l,m,n} k_{lmn}\sum_{G} D^{\gamma}_{ll'}(G)w^{\gamma}_{l'}\, D^{\mu}_{mm'}(G)u^{\mu}_{m'}\, D^{\nu}_{nn'}(G)v^{\nu}_{n'} \sim \mathbf{1}. \tag{5}$$

Clearly, the method described above can be applied to any finite group. In concrete cases, however, there may exist a more practical way to obtain tensor products and decompose them into irreps. This is exactly the case for the dihedral groups as we will see later on.

2.3. *Finite groups of lower orders*

There are several facts which seem to be relevant in applying finite groups of lower orders as a family group of elementary particles.

- There exists no non-abelian finite group for odd g.

- For smaller g, i.e. $g < 37$, there exist only three types of non-abelian finite groups:[9]
  - (a) Even and odd permutations $S_N, N = 3, 4, 5, \dots,$ and Even permutations $A_N, N = 4, 5, 6, \dots$.
  - (b) Dihedral groups: $D_N, N = 3, 4, 5, \dots,$ and binary dihedral (dicyclic) groups $Q_N, N = 4, 6, 8, \dots$.
  - (c) Twisted products of abelian groups, i.e. $Z_M \tilde{\times}\, Z_N(\text{or}D_N)$ with $[Z_M, Z_N(\text{or}D_N)] \neq 0$.
- The smallest non-abelian group is $S_3 \sim D_3$.

3. The dihedral group

$D_N(N = 3, 4, 5, \dots)$ is a group which can be generated by R_N and P_D

$$\{\ R_N, P_D; (R_N)^N = P_D^2 = \mathbf{1}, P_D R_N P_D^{-1} = R_N^{-1}\ \}, \tag{6}$$

and similarly for $Q_N(N = 4, 6, 8, \dots)$,

$$\{\ R_N, P_Q; (R_N)^N = \mathbf{1}, (R_N)^{N/2} = P_Q^2, P_Q R_N P_Q^{-1} = R_N^{-1}\ \}. \tag{7}$$

Then the $2N$ group elements are given by

$$\begin{aligned}\mathcal{G}_{D_N(Q_N)} = \{&\mathbf{1}, R_N, (R_N)^2, \dots, (R_N)^{N-1}, P_{D(Q)},\\ &R_N P_{D(Q)}, (R_N)^2 P_{D(Q)}, \dots, (R_N)^{N-1} P_{D(Q)}\}.\end{aligned} \tag{8}$$

It is straightforward to show that $\mathcal{G}_{D_N(Q_N)}$ forms a group.

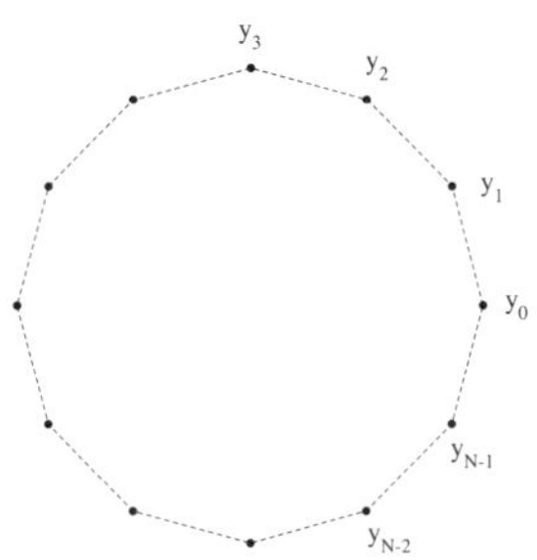

Fig. 1. A regular polygon with $N = 12$ edges, which are located at $y = y_0, y_1, \dots, y_{N-1}$.

3.1. *Matrix presentations of D_N and Q_{2N}*

The dihedral group D_N is a symmetry group of the regular polygon with N edges, which is plotted in Fig. 1 for $N = 12$. The D_N operations are $2N$ discrete rotations, where N of $2N$ rotations are combined with a parity transformation. The angle of the fundamental discrete polygon rotation is given by

$$\theta_N \equiv 2\pi/N. \tag{9}$$

The N sites are located at $y = y_0, y_1, \ldots, y_{N-1}$. ($y_{N+i}$ is identified with y_i.) Under a D_N transformation, the set of coordinates $(y_0, y_1, \ldots, y_{N-1})$ changes to $(y'_0, y'_1, \ldots, y'_{N-1})$, which we express in terms of a $N \times N$ real matrix. The matrix for the fundamental rotation is given by

$$\tilde{R}_N = \begin{pmatrix} 0 & 0 & 0 & \cdots & 1 \\ 1 & 0 & 0 & \cdots & 0 \\ 0 & 1 & 0 & \cdots & 0 \\ & & \cdots & & \\ 0 & 0 & \cdots & 1 & 0 \end{pmatrix}, \tag{10}$$

and that for the parity transformation is

$$\tilde{P}_D = \begin{pmatrix} 1 & 0 & \cdots & 0 & 0 \\ 0 & \cdots & 0 & 0 & 1 \\ 0 & \cdots & 0 & 1 & 0 \\ & & \cdots & & \\ 0 & 1 & 0 & \cdots & 0 \end{pmatrix}. \tag{11}$$

Then one can easily find that $\tilde{P}_D^2 = \mathbf{1}$ and $\tilde{P}_D \tilde{R}_N \tilde{P}_D^{-1} = (R_N)^{-1}$, so that $\tilde{R}_N$ and $\tilde{P}_D$ can be used to obtain the regular representations of D_N.

There exist two-dimensional presentation for R_N and P_D[2,9]

$$R_N = \begin{pmatrix} \cos\theta_N & \sin\theta_N \\ -\sin\theta_N & \cos\theta_N \end{pmatrix}, \quad P_D = \begin{pmatrix} 1 & 0 \\ 0 & -1 \end{pmatrix}, \tag{12}$$

which satisfy $P_D R_N P_D = R_N^{-1}$. It follows that D_N is a subgroup of $SO(3)$, which one sees if one embeds R_N and P_D into 3×3 matrices

$$R_N \to \begin{pmatrix} \cos\theta_N & \sin\theta_N & 0 \\ -\sin\theta_N & \cos\theta_N & 0 \\ 0 & 0 & 1 \end{pmatrix}, P_D \to \begin{pmatrix} 1 & 0 & 0 \\ 0 & -1 & 0 \\ 0 & 0 & -1 \end{pmatrix}. \tag{13}$$

As in the case of $SO(3)$, all the representations of D_N are real.

$SU(2)$ is the universal covering group of $SO(3)$, and has pseudo real and real irreps. All the real irreps of $SU(2)$ are those of $SO(3)$. Q_{2N} is a finite subgroup of $SU(2)$, and in a similar sense, Q_{2N} can be regarded as the covering group of D_N. The two-dimensional matrices generating Q_N are:

$$R_N = \begin{pmatrix} \cos\theta_N & \sin\theta_N \\ -\sin\theta_N & \cos\theta_N \end{pmatrix}, \; P_Q = \begin{pmatrix} i & 0 \\ 0 & -i \end{pmatrix}, \tag{14}$$

where $(R_N)^{N/2} = (P_Q)^2 = -\mathbf{1}$, and θ_N is defined in (9).

3.2. *Irreducible representations of D_N and Q_N*

Before we proceed, consider an $SO(2)$ vector

$$\mathbf{x} = \begin{pmatrix} x \\ y \end{pmatrix}, \; x, y \in \mathbf{C}. \tag{15}$$

Under an $SO(2)$ rotation, the vector transforms as

$$\mathbf{x} \to \mathbf{x}' = \begin{pmatrix} x' \\ y' \end{pmatrix} = \begin{pmatrix} \cos\theta & \sin\theta \\ -\sin\theta & \cos\theta \end{pmatrix} \begin{pmatrix} x \\ y \end{pmatrix} = \begin{pmatrix} x\cos\theta + y\sin\theta \\ -x\sin\theta + y\cos\theta \end{pmatrix}.$$

Then we go to the $U(1)$ notation. Define

$$z = x + iy, \tag{16}$$

which transforms as

$$z \to z' = x' + iy' = e^{-i\theta}(x + iy) = e^{-i\theta} z. \tag{17}$$

From (17) we obtain

$$\hat{z} = \hat{x} + i\hat{y} = -y + ix \to \hat{z}' = \hat{x}' + i\hat{y}' = e^{-i\theta}(-y + ix) = e^{-i\theta}\hat{z}, \tag{18}$$

$$\overline{z} = \overline{x} + i\overline{y} = x - iy \to \overline{z}' = \overline{x}' + i\overline{y}' = e^{i\theta}\overline{z}. \tag{19}$$

<u>D_N</u>

With these remarks, we consider a D_N vector $\mathbf{2} = (x, y)^T$, which transform as

$$z = x + iy \to z' = x' + iy' = e^{-i\theta_N} z, \tag{20}$$

where $\theta_N = 2\pi/N$ is the angle of the fundamental rotation of D_N. Note that because $(R_N)^N = \mathbf{1}$, we have $R_N \in Z_N$. Therefore, we assign the vector $\mathbf{2}$ the Z_N charge of one, and so we write it as

$$\mathbf{2}_1 : z(1) = x(1) + iy(1) \tag{21}$$

in the $U(1)$ notation. There are N different charges (irreps) of Z_N: $0, 1, 2, \ldots, N-1$, where the irrep with zero charge is the singlet. Therefore, one may expect that there exist $N-2$ two-dimensional irreps for D_N. But this is not the case; there exits only $(N-2)/2$ for an even N, and $(N-1)/2$ for an odd N. The reason is the following. Consider a vector $\mathbf{2}_m = (x(m), y(m))^T$ (whose Z_N charge is m), and

$$\overline{\mathbf{x}}(m) = \begin{pmatrix} \overline{x}(m) \\ \overline{y}(m) \end{pmatrix} = P_D \begin{pmatrix} x(m) \\ y(m) \end{pmatrix} = \begin{pmatrix} x(m) \\ -y(m) \end{pmatrix}, \tag{22}$$

where P_D is defined in (12). The Z_N charge of $\overline{\mathbf{x}}(m)$ is obviously $-m$. Therefore, under the party transformation P_D, the states with $Z_N = m$ go over to the states with $Z_N = -m$; they belong to the same irrep $\mathbf{2}_m$. So, for an odd N, there exist $(N-1)/2$ two-different irreps. For an even N, there exist $N/2-1$, because $\mathbf{2}_{N/2}$ transforms as

$$z(N/2) \to z'(N/2) = e^{-i(2\pi/N)(N/2)} z(N/2) = -z(N/2), \tag{23}$$

which means that

$$x(N/2) \text{ and } y(N/2) \text{ are singlets (one-dimensional irreps).} \tag{24}$$

For these singlets, R_N acts as Z_2, where their Z_2 parity is odd. Note that P_D is also an element of Z_2, so that the all singlets for even N can be characterized according to $Z_2 \times P_D$ parity:

$$\begin{aligned} &\mathbf{1}_{++},\ \mathbf{1}_{-+},\ \mathbf{1}_{+-},\ \mathbf{1}_{--} \quad \text{for even } N, \\ &\mathbf{1}_{++},\ \mathbf{1}_{+-} \quad \text{for odd } N, \end{aligned} \tag{25}$$

where the $\mathbf{1}_{++}$ is the true singlet of D_N.

Q_N

The same discussions above can be applied to Q_N, so that there exist $N/2-2$ two-dimensional irreps. (N is always even for Q_N.) However, the irreps of Q_N can be complex, because P_Q is a complex matrix (see (14)). To see this, consider a Q_N vector $\mathbf{2}_1 = (x(1), y(1))^T$ with Z_N charge one. Under P_Q it transforms as

$$z(1) = x(1) + iy(1) \to z'(1) = x'(1) + iy'(1) = ix(1) + y(1) = i\overline{z}, \tag{26}$$

where $\overline{z}$ is defined in (19). On the other hand, $z^*(1)$ transforms as

$$\begin{aligned} z^*(1) = x^*(1) - iy^*(1) \to z'^*(1) &= x'^*(1) - iy'^*(1) \\ &= -ix^*(1) + y^*(1) = -i\overline{z^*}, \end{aligned} \tag{27}$$

implying that $z(1)$ is a complex irrep. Similarly, the singlets given in (25) can be complex irreps for Q_N. To see this, consider again a Q_N vector $\mathbf{2}_1$:

$$z(1) \to z'(1) = e^{-i(2\pi/N)} z(1) \quad \text{under } R_N. \tag{28}$$

Under P_Q, $z(1)$ transforms as

$$z(1) \to z'(1) = x'(1) + iy'(1) = ix(1) + y(1). \tag{29}$$

Then consider a product of two vectors $z_1(1)z_2(1)$:

$$\begin{aligned} z_1(1)z_2(1) &= z_3(2) = x_3(2) + iy_3(2) \\ &= x_1(1)x_2(1) - y_1(1)y_2(1) + i[x_1(1)y_2(1) + y_1(1)x_2(1)] \\ &\to z_3'(2) = x_3'(2) + iy_3'(2) \\ &\qquad = -x_1(1)x_2(1) + y_1(1)y_2(1) + i[x_1(1)y_2(1) + y_1(1)x_2(1)] \\ &\qquad = -x_3(2) + iy_3(2). \end{aligned} \tag{30}$$

Using (18), we then obtain

$$\begin{aligned} \hat{z}_3(2) &= \hat{x}_3(2) + i\hat{y}_3(2) = -y_3(2) + ix_3(2) \\ &\to \hat{z}_3'(2) = \hat{x}_3'(2) + i\hat{y}_3'(2) = -y_3'(2) + ix_3'(2) \\ &= -[x_1(1)y_2(1) + y_1(1)x_2(1)] + i[-x_1(1)x_2(1) + y_1(1)y_2(1)] \\ &= -y_3(2) - ix_3(2) = \hat{x}_3(2) - i\hat{y}_3(2). \end{aligned} \tag{31}$$

This means that $\hat{z}_3(2)$ transforms like $z_1(1)$ of $D_{N/2}$, because $\hat{z}_3(2)$ of Q_N rotates with $(2\pi/N)2 = 2\pi/(N/2)$. Therefore, the parity transformation for $\hat{z}_3(2)$ is presented by P_D and moreover it is a real representation, because under P_D

$$x_3(2), y_3(2) \to x_3(2), -y_3(2) \quad \text{and} \quad x_3^*(2), y_3^*(2) \to x_3^*(2), -y_3^*(2). \tag{32}$$

With these remarks, we then consider singlets which can be obtained from

$$z_1(1)z_2(1)\ldots z_{N/2}(1). \tag{33}$$

(33) contains two singlet irreps; the real and imaginary parts. Clearly, (33) is a product of $N/2$ complex irreps, so that if $N/2$ is even, the singlets are real irreps and they are complex irreps for odd $N/2$. They can be characterized according to $Z_2 \times Z_4$ charge, because $P_Q \in Z_4$:

$$\mathbf{1}_{+0},\ \mathbf{1}_{-0},\ \mathbf{1}_{+2},\ \mathbf{1}_{-2} \quad \text{for} \quad N = 4, 8, 12, \ldots, \tag{34}$$

$$\mathbf{1}_{+0},\ \mathbf{1}_{-1},\ \mathbf{1}_{+2},\ \mathbf{1}_{-3} \quad \text{for} \quad N = 6, 10, 14, \ldots, \tag{35}$$

where the $\mathbf{1}_{+0}$ is the true singlet of Q_N, and only $\mathbf{1}_{-1}$ and $\mathbf{1}_{-3}$ are complex irreps.

To summarize, there exist only one- and two-dimensional irreps for D_N and Q_N. For Q_N, there are $N/2-1$ different two-dimensional irreps, which we denote by $\mathbf{2}_\ell$, $\ell = 1,\ldots,N/2-1$. $\mathbf{2}_\ell$ with odd ℓ is a pseudo real representation, while $\mathbf{2}_\ell$ with even ℓ is a real representation. The singlets can be characterized according to $Z_2 \times P_D$ for D_N, and to $Z_2 \times Z_4$ for Q_N. All the real singlets of Q_N are those of $D_{N/2}$:

$$\mathbf{1}_{+0} = \mathbf{1}_{++},\ \mathbf{1}_{-0} = \mathbf{1}_{-+},\ \mathbf{1}_{+2} = \mathbf{1}_{+-},\ \mathbf{1}_{-2} = \mathbf{1}_{--} \quad \text{for} \quad N/2 = 4,6,8,\ldots, \tag{36}$$

$$\mathbf{1}_{+0} = \mathbf{1}_{++},\ \mathbf{1}_{+2} = \mathbf{1}_{+-} \quad \text{for} \quad N/2 = 3,5,7,\ldots, \tag{37}$$

$\mathbf{2}_\ell$ of Q_N with even ℓ is exactly $\mathbf{2}_{\ell/2}$ of $D_{N/2}$. So, all real irreps of of Q_N are those of $D_{N/2}$, which is the reason why we would like to call Q_N as the covering group of $D_{N/2}$.

3.3. *Tensor products*

Making a tensor product of two irreps is basically an addition of two Z_N charges. There are only four types of $\mathbf{2}\otimes\mathbf{2}$:

$$\mathbf{2}\otimes\mathbf{2} = \begin{cases} \text{four } \mathbf{1}'\text{s} & \text{Type A} \\ \text{two } \mathbf{2}'\text{s} & \text{Type B}\,. \\ \text{two } \mathbf{1}'\text{s} \oplus \text{one } \mathbf{2} & \text{Type C} \end{cases} \tag{38}$$

Type A

Consider the product $z_1(m_1)z_2(m_2)$ of D_N or Q_N with an even N. $z_1(m_1)z_2(m_2) = z_3(m_1+m_2)$ contains two singlets if

$$|m_1 + m_2| = 0 \text{ or } N/2. \tag{39}$$

Similarly, $z_1(m_1)\overline{z}_2(m_2) = z_3(m_1 - m_2)$ ($\overline{z}$ is defined in (19)) contains two singlets if

$$|m_1 - m_2| = 0 \text{ or } N/2. \tag{40}$$

Therefore, there are two cases:

$$A_1 : |m_1 + m_2| = 0 \text{ and } |m_1 - m_2| = N/2, \tag{41}$$

$$A_2 : |m_1 + m_2| = N/2 \text{ and } |m_1 - m_2| = 0. \tag{42}$$

For both cases one obtains $|m_1| = |m_2| = N/4$. This means that the type A can appear only if N is a multiple of 4. In fact, for D_4 and Q_4, there are only this type of tensor product of doublets.

Type B

Consider again the product $z_1(m_1)$ and $z_2(m_2)$ of D_N or Q_N, where we assume $m_1, m_2 > 0$. There are two types of products:

$$z_3(m_1 + m_2) = z_1(m_1)z_2(m_2) \text{ and } z_3(m_1 - m_2) = z_1(m_1)\overline{z}_2(m_2). \quad (43)$$

$z_3(m_1 + m_2)$ corresponds to

$$\begin{pmatrix} x_1(m_1)x_2(m_2) - y_1(m_1)y_2(m_2) \\ x_1(m_1)y_2(m_2) + y_1(m_1)x_2(m_2) \end{pmatrix}. \quad (44)$$

The Z_N charge of $z_3(m_1 + m_2)$ is $m_1 + m_2$, so that if $m_1 + m_2 > N/2$, we use $\overline{z}_3(m_1 + m_2)$ whose Z_N charge is $N - (m_1 + m_2)$, and corresponds to

$$\begin{pmatrix} x_1(m_1)x_2(m_2) - y_1(m_1)y_2(m_2) \\ -x_1(m_1)y_2(m_2) - y_1(m_1)x_2(m_2) \end{pmatrix}. \quad (45)$$

For D_N, these are the only possibilities for the product of $z_1(m_1)$ and $z_2(m_2)$, because the transformation property under P_D turns out to be always correct. As for Q_N, this is not the case. If the transformation under P_Q turns out to be upside down, then we have to use $\hat{z}$. Then (44) and (45), respectively, change to

$$\begin{pmatrix} -x_1(m_1)y_2(m_2) - y_1(m_1)x_2(m_2) \\ x_1(m_1)x_2(m_2) - y_1(m_1)y_2(m_2) \end{pmatrix}, \begin{pmatrix} x_1(m_1)y_2(m_2) + y_1(m_1)x_2(m_2) \\ x_1(m_1)x_2(m_2) - y_1(m_1)y_2(m_2) \end{pmatrix}. \quad (46)$$

(44),(45) and (46) can be written in a more compact form:

$$\mathbf{2}_{sN+(-1)^s(m_1+m2)} = \begin{pmatrix} 0 & -1 \\ 1 & 0 \end{pmatrix}^t \begin{pmatrix} x_1(m_1)x_2(m_2) - y_1(m_1)y_2(m_2) \\ (-1)^s[x_1(m_1)y_2(m_2) + y_1(m_1)x_2(m_2)] \end{pmatrix},$$

$$s = \begin{cases} 0 \\ 1 \end{cases} \text{for} \begin{cases} m_1 + m_2 < N/2 \\ m_1 + m_2 > N/2 \end{cases}, \quad t = \begin{cases} 0 \\ 1 \end{cases} \text{if } P_Q \text{ is} \begin{cases} \text{correct} \\ \text{upside down} \end{cases}. \quad (47)$$

Similarly, for the second product of (43) we obtain:

$$\mathbf{2}_{m_1-m2} = \begin{pmatrix} 0 & -1 \\ 1 & 0 \end{pmatrix}^t \begin{pmatrix} x_1(m_1)x_2(m_2) + y_1(m_1)y_2(m_2) \\ -x_1(m_1)y_2(m_2) + y_1(m_1)x_2(m_2) \end{pmatrix}, \quad (48)$$

$$t = \begin{cases} 0 \\ 1 \end{cases} \text{if } P_Q \text{ is} \begin{cases} \text{correct} \\ \text{upside down} \end{cases},$$

where $m_1 > m_2 > 0$ is assumed without loss of generality.

From (47) and (48) we find that the condition that there exist two doublets in $\mathbf{2} \otimes \mathbf{2}$ is that

$$|m_1 + m_2|, |m_1 - m_2| \neq 0 \text{ nor } N/2. \tag{49}$$

Type C

From the discussion of the type B, it is clear that if

$$|m_1 + m_2| \neq 0 \text{ nor } N/2 \text{ and } |m_1 - m_2| = 0 \text{ or } N/2, \tag{50}$$

OR

$$|m_1 - m_2| \neq 0 \text{ nor } N/2 \text{ and } |m_1 + m_2| = 0 \text{ or } N/2 \tag{51}$$

is satisfied, there will be one doublet and two singlets in $\mathbf{2} \otimes \mathbf{2}$.

3.4. *Q_6 as an example*

The smallest group that contains real and pseudo real irreps is Q_6, which is the double-covering group of $S_3 \sim D_3$. The irreps of Q_6 with $N/2 = 3$ are $\mathbf{2}_1$, $\mathbf{2}_2$, $\mathbf{1}_{+0}$, $\mathbf{1}_{-1}$, $\mathbf{1}_{+2}$, $\mathbf{1}_{-3}$, where the $\mathbf{2}_1$ is pseudo-real, while $\mathbf{2}_2$ is real. $\mathbf{1}_{+0}$, $\mathbf{1}_{+2}$ are real representations, while $\mathbf{1}_{-1}$, $\mathbf{1}_{-3}$ are complex conjugate to each other.

First we consider the tensor product $\mathbf{2}_1 \otimes \mathbf{2}_1$ or $z_1(1) \otimes z_2(1)$. According to the section 3.3, the product corresponds to the type C; two $\mathbf{1}$'s and one $\mathbf{2}$. Since $m_1 = m_2 = 1$ so that $m_1 - m_2 = 0$, the two singlets can be obtained from $z_1(1)\bar{z}_2(1) = x_1(1)x_2(1) + y_1(1)y_2(1) + i[-x_1(1)y_2(1) + y_1(1)x_2(1)]$:

$$\mathbf{1}_{+0}(\subset \mathbf{2}_1 \otimes \mathbf{2}_1) = -x_1(1)y_2(1) + y_1(1)x_2(1), \tag{52}$$
$$\mathbf{1}_{+2}(\subset \mathbf{2}_1 \otimes \mathbf{2}_1) = x_1(1)x_2(1) + y_1(1)y_2(1). \tag{53}$$

To obtain $\mathbf{2}_2$ we then consider $z_1(1)z_2(1) = x_1(1)x_2(1) - y_1(1)y_2(1) + i[x_1(1)y_2(1) + y_1(1)x_2(1)]$. Since $m_1 + m_2 = 2 < 6/2$, we use the case (47) with $s = 0$. Then $t = 1$ ensures the correct transformation under P_D:

$$\mathbf{2}_2(\subset \mathbf{2}_1 \otimes \mathbf{2}_1) = \begin{pmatrix} -x_1(1)y_2(1) - y_1(1)x_2(1) \\ x_1(1)x_2(1) - y_1(1)y_2(1) \end{pmatrix}, \tag{54}$$

which has the correct transformation property under P_D.

Next we consider the tensor product $\mathbf{2}_1 \otimes \mathbf{2}_2$ or $z_1(1) \otimes z_2(2)$. The product corresponds to the type C, too. Since $m_1 = 1, m_2 = 2$ so that $m_1 + m_2 = 3 = N/2$, the two singlets can be obtained from $z_1(1)z_2(2) = x_1(1)x_2(2) -$

$y_1(1)y_2(2) + i[x_1(1)y_2(2) + y_1(1)x_2(2)]$:

$$\mathbf{1}_{-3}(\subset \mathbf{2}_1 \otimes \mathbf{2}_2) = x_1(1)y_2(2) + y_1(1)x_2(2), \tag{55}$$

$$\mathbf{1}_{-1}(\subset \mathbf{2}_1 \otimes \mathbf{2}_2) = x_1(1)x_2(2) - y_1(1)y_2(2). \tag{56}$$

To obtain $\mathbf{2}_1$ we consider $\overline{z}_1(1)z_2(2) = x_1(1)x_2(2) + y_1(1)y_2(2) + i[x_1(1)y_2(2) - y_1(1)x_2(2)]$. (48) with $t = 0$ gives

$$\mathbf{2}_1(\subset \mathbf{2}_1 \otimes \mathbf{2}_2) = \begin{pmatrix} x_1(1)x_2(2) + y_1(1)y_2(2) \\ x_1(1)y_2(2) - y_1(1)x_2(2) \end{pmatrix}, \tag{57}$$

which has the correct transformation property under P_Q.

Similarly we find for $\subset \mathbf{2}_2 \otimes \mathbf{2}_2$:

$$\mathbf{1}_{+2}(\subset \mathbf{2}_2 \otimes \mathbf{2}_2) = -x_1(2)y_2(2) + y_1(2)x_2(2), \tag{58}$$

$$\mathbf{1}_{+0}(\subset \mathbf{2}_2 \otimes \mathbf{2}_2) = x_1(2)x_2(2) + y_1(2)y_2(2). \tag{59}$$

To obtain $\mathbf{2}$ we consider $z_1(2)z_2(2) = x_1(2)x_2(2) - y_1(2)y_2(2) + i[x_1(2)y_2(2) + y_1(2)x_2(2)]$. Since $m_1 + m_2 = 4 > 6/2$, we use the case (47) with $s = 1$ and $t = 0$. Then we obtain

$$\mathbf{2}_2(\subset \mathbf{2}_2 \otimes \mathbf{2}_2) = \begin{pmatrix} x_1(2)x_2(2) - y_1(2)y_2(2) \\ -x_1(2)y_2(2) - y_1(2)x_2(2) \end{pmatrix}. \tag{60}$$

The results obtained above are used in applying Q_6 as a family group, which will follow below.

4. The model

As announced in the introduction, we are here interested in flavor models with a low energy family symmetry that make predictions testable at future B factories. To our knowledge[3] there exists at present only one model[2] which possesses the following properties: (i) The family symmetry is a symmetry both in the quark and lepton sectors, making predictions in both sectors. That is, in the quark sector, the model describes 10 observables, i.e. six quark masses and four CKM parameters, by less than 10 parameters, and in the lepton sector, it describes 12 observables, i.e. three charged lepton masses, three neutrino masses and six Maki-Nakagawa-Sakata (MNS) parameters, by less than 12 parameters. (ii) The family symmetry is not hardly broken; it is broken softly at most. (iii) The model is within the frame work of renormalizability.

The $SU(2)_L$ doublets of the quark and Higgs supermultiplets are denoted by Q and H^u, H^d, respectively. (Here we restrict ourselves to the quark sector. See[10,11] for a Q_6 assignment of the leptons to obtain the

Table 1. The Q_6 assignment of the matter multiplets. The multiplication rules can be found in 3.4.

	$Q_{1,2}$	$U^c_{1,2}\ D^c_{1,2}$	$H^u_{1,2}\ H^d_{1,2}$	Q_3	$U^c_3\ D^c_3$	$H^u_3\ H^d_3$
Q_6	$\mathbf{2}_1$	$\mathbf{2}_2$	$\mathbf{2}_2$	$\mathbf{1}_{+2}$	$\mathbf{1}_{-1}$	$\mathbf{1}_{-1}$

maximal mixing of the atmospheric neutrinos. The prediction in this sector is exactly the same as the S_3 model of.[12]) Similarly, $SU(2)_L$ singlets of the quark supermultiplets are denoted by U^c and D^c. As we have seen in section 3.4, the finite group Q_6 allows complex representations, and the Q_6 assignment of the matter multiplets is given in Table 1.

The superpotential for the Yukawa interactions in the quark sector is given by

$$\begin{aligned} W_q &= Y_a^u Q_3 H_3^u U_3^c + Y_b^u (Q_1 H_2^u + Q_2 H_1^u) U_3^c + Y_c^u Q_3 (H_1^u U_2^c - H_2^u U_1^c) \\ &\quad + Y_e^u (Q_1 U_2^c + Q_2 U_1^c) H_3^u \\ &\quad + Y_a^d Q_3 H_3^d D_3^c + Y_b^d (Q_1 H_2^d + Q_2 H_1^d) D_3^c + Y_c^d Q_3 (H_1^d D_2^c - H_2^d D_1^c) \\ &\quad + Y_e^d (Q_1 D_2^c + Q_2 D_1^c) H_3^d \ . \end{aligned} \tag{61}$$

To make the model predictive there are two crucial requirements: (1) the VEV alignment

$$< H_1^{u,d} > = < H_2^{u,d} > \equiv v_1^{u,d} \ , \ < H_3^{u,d} > \equiv v_3^{u,d}, \tag{62}$$

which can be achieved by an accidental permutation symmetry $H_1^{u,d} \leftrightarrow H_2^{u,d}$ in the Higgs sector, and (2) CP is spontaneously broken. The second requirement can be relaxed to that the Yukawa couplings are real without contradicting renormalizability [b]. Then the quark mass matrices can be written as

$$M_u = \begin{pmatrix} 0 & Y_e^u\ v_3^u & Y_b^u\ v_1^u \\ Y_e^u\ v_3^u & 0 & Y_b^u\ v_1^u \\ -Y_c^u\ v_1^u & Y_c^u\ v_1^u & Y_a^u\ v_3^u \end{pmatrix}, M_d = \begin{pmatrix} 0 & Y_e^d\ v_3^d & Y_b^d\ v_1^d \\ Y_e^d\ v_3^d & 0 & Y_b^d\ v_1^d \\ -Y_c^d\ v_1^d & Y_c^d\ v_1^d & Y_a^d\ v_3^d \end{pmatrix} \tag{63}$$

with complex VEVs.

By making an overall 45° rotation of the Q_6 doublets Q, U^c and D^c in the space of the family group, we obtain nearest neighbor interaction type

[b] It has been found[11] that to trigger complex VEVs with the minimal content of the chiral supermultiplets given in Table 1, the family symmetry and CP should be softly broken at least by certain dimension two operators in the soft-supersymmetry breaking sector.

mass matrices:

$$M_u = \begin{pmatrix} 0 & c_u & 0 \\ -c_u & 0 & b_u \\ 0 & d_u & e_u \end{pmatrix}, M_d = \begin{pmatrix} 0 & c_d & 0 \\ -c_d & 0 & b_d \\ 0 & d_d & e_d \end{pmatrix} . \tag{64}$$

All the elements of these matrices can be made real by a redefinition of the quark fields. Then the real matrices can be diagonalized by orthogonal matrices as

$$O_u^T M_u M_u^T O_u = \text{diag}(m_u^2, m_c^2, m_t^2), \tag{65}$$

$$O_d^T M_d M_d^T O_d = \text{diag}(m_d^2, m_s^2, m_b^2), \tag{66}$$

and the CKM matrix takes the form $V_{\text{CKM}} = O_u^T P O_d$, where $P = \text{diag}(1, e^{2i\theta}, e^{i\theta})$. The phase rotation matrix P has only one angle θ, which is the consequence of a spontaneously, softly broken CP.

4.1. *Predictions*

According to,[13] the CKM matrix can be approximately written in a closed form. One finds, for instance,

$$V_{us} \simeq -y_d \sqrt{\frac{m_d}{m_s}} + y_u \sqrt{\frac{m_u}{m_c}}\ e^{2i\theta} , \tag{67}$$

$$V_{cb} \simeq \frac{y_d^2}{\sqrt{1-y_d^4}} \frac{m_s}{m_b}\ e^{2i\theta} - \frac{y_u^2}{\sqrt{1-y_u^4}} \frac{m_c}{m_t}\ e^{i\theta}, \tag{68}$$

$$\begin{aligned} V_{ub} \simeq & \frac{\sqrt{1-y_d^4}}{y_d} \sqrt{\frac{m_d}{m_s}\frac{m_s}{m_b}} \\ & + y_u \sqrt{\frac{m_u}{m_c}} \left(\frac{y_d^2}{\sqrt{1-y_d^4}} \frac{m_s}{m_b}\ e^{2i\theta} - \frac{1}{y_u^2\sqrt{1-y_u^4}} \frac{m_c}{m_t}\ e^{i\theta} \right) , \end{aligned} \tag{69}$$

where $y_u = 1/c_u$ and $y_d = 1/c_d$. For $y_d \simeq 1$ we can reproduce the classic relation of[14–16]

$$|V_{us}| \simeq \sqrt{\frac{m_d}{m_s}} + O(m_u/m_c). \tag{70}$$

We have performed numerical analyses on the predictions of the model in detail,[3] some of which are presented in Figs. 2 - 5. The experimental values of the CKM parameters are also included: $\bar{\rho} = 0.239 \pm 0.046$, $\bar{\eta} = 0.326 \pm 0.031, \phi_2(\alpha) = 101.5 \pm 10.5$, $\phi_3(\gamma) = 64.5 \pm 13.5$, $\sin 2\phi_1(\beta) = 0.687 \pm 0.032$ (black),[17] and $\bar{\rho} = 0.147 \pm 0.029$, $\bar{\eta} = 0.342 \pm 0.012, \phi_2(\alpha) = 91.2 \pm 6.1$, $\phi_3(\gamma) = 66.7 \pm 6.4$, $\sin 2\phi_1(\beta) = 0.690 \pm 0.023$, $|V_{td}/V_{ts}| = 0.211 \pm 0.007$ (blue).[18] The model is so far consistent with the PDG values[17] as well as with the Utfit group values.[18] At $\phi_3 \simeq 65^0$, the band of the predicted ϕ_2 is about 20^0 wide. We found that this wide band mainly originates from the uncertainty of the strange quark mass m_s. In Fig. 5 we plot two predicted regions for two different sets of the mass values of m_s and m_d. We see that precise measurements of ϕ_2 can distinguish the two regions. We also found that the uncertainty in the strange quark mass should be less than few % to make it comparable with the assumed uncertainties of $\sim 1^0$ and $\sim 2^0$ in ϕ_2 and ϕ_3, respectively, at about 50 inverse atto barn which may be achieved at a feature B factory.

5. Zum Scluss

und nicht zuletzt möchte ich die heutige Gelegenheit nutzen, um mich bei Professor Zimmermann für die langjärige Unterstützung zu bedanken. Am 1. Januar 1984 fing meine Arbeit als Postdoc am MPI an. Ich hatte gleich das Glück, Klaus Sibold kennenzulernen, weil ich durch ihn zur Zusammenarbeit mit Professor Zimmermann kam. Ich blieb für anderthalb Jahre am MPI und wechselte dann nach Stony Brook. In Stony Brook bekam ich den ersten Brief von Professor Zimmermann und entschloss mich ihn nicht zu verlieren. Nicht nur den ersten, sondern alle weiteren, habe ich dann behalten. Wie diese vielen Briefe zeigen, hat mich Professor Zimmermann seit 24 Jahren ständig unterstützt. Ich danke Ihnen oder ihm von Herzen. Allerdings waren diese Briefe nicht ohne Probleme. Es war für mich nicht einfach, seine Briefe zu lesen. Manchmal musste ich mit meiner Frau lange zusammensitzen, um die Briefe vollstädig zu analisieren.

Zum Fruehlingsanfang 1977 besuchte uns Professor Zimmermann für eine längere Zeit. Er genoss die schöne japanische Kirschblüte. Wir alle wissen, dass er gerne isst. In Japan hat er auch verschiedenen Dinge probiert. Unter anderen die Tsumamis. Sie sind Knabbereien, die man zum Trinken in Japan dazu isst. Oft sind es getrocknete Fische. Ich habe ihm heute Tsumamis mitgebracht. Ich hoffe, dass er sie noch mag. Noch etwas habe ich ihm mitgebracht, das die meisten Ausländer in Japan nicht mögen. Das ist Natto. Natto besteht aus fermentierten (verdorbenen) Sojabohnen. Obwohl meine Frau in den ersten 15 Jahren in Japan Natto nicht essen konnte, konnte Professor Zimmermann Natto gleich essen. Sogar meinte er, sie schmeckten gut.

Ich wïsche Professor Zimmermann noch viele glückliche Jahre mit seiner Frau, den Kindern und Enkelkindern und auch guten Appetit.

Zum Schluss bedanke ich mich bei Erhard, Klaus und Rosvita für ihre grosse Mühe dieses netten, ausserordentlichen Treffens. Es war eine Ehre für mich zum Anlass des 80sten Geburtstages von Professor Zimmermann einen Vortrag halten zu dürfen.

References

1. S. Pakvasa and H. Sugawara, *Phys. Lett. B* **73**, 61 (1978).
2. K. S. Babu and J. Kubo, *Phys. Rev. D* **71**, 056006 (2005) [arXiv:hep-ph/0411226].
3. T. Araki, M. Hazumi and J. Kubo, to appear.
4. A. G. Akeroyd *et al.* [SuperKEKB Physics Working Group], arXiv:hep-ex/0406071.
5. T. Browder, M. Ciuchini, T. Gershon, M. Hazumi, T. Hurth, Y. Okada and A. Stocchi, arXiv:0710.3799 [hep-ph].
6. S. Okubo, *Phys. Rev. D* **12**, 3835 (1975).
7. L. D. Landau and E. M. Lifshitz, *Quantum Mechanics: Non-Relativistic Theory* (Pergamon Press, Oxford,1976).
8. Wu-Ki Tung, *Group Theory in Physics* (World Scientific, Singapore, 1985).
9. P. H. Frampton and T. W. Kephart, *Int. J. Mod. Phys. A* **10**, 4689 (1995); *Phys. Rev. D* **64**, 086007 (2001).
10. E. Itou, Y. Kajiyama and J. Kubo, *Nucl. Phys. B* **743**, 74 (2006).
11. N. Kifune, J. Kubo and A. Lenz, *Phys. Rev. D* **77**, 076010 (2008).
12. J. Kubo, A. Mondragon, M. Mondragon and E. Rodriguez-Jauregui, *Prog. Theor. Phys.* **109**, 795 (2003) [Erratum-ibid. **114**, 287 (2005)].
13. K. Harayama, N. Okamura, A. I. Sanda and Z. Z. Xing, *Prog. Theor. Phys.* **97**, 781 (1997).
14. R. Gatto, G. Sartori and M. Tonin, *Phys. Lett. B* **28**, 128 (1968).
15. N. Cabibbo and L. Maiani, *Phys. Lett. B* **28**, 131 (1968).
16. H. Fritzsch, *Phys. Lett. B* **70**, 436 (1977).
17. W. M. Yao *et al.* [Particle Data Group], *J. Phys. G* **33**, 1 (2006).
18. M. Bona *et al.* [UTfit Collabaration], http://www.utfit.org/

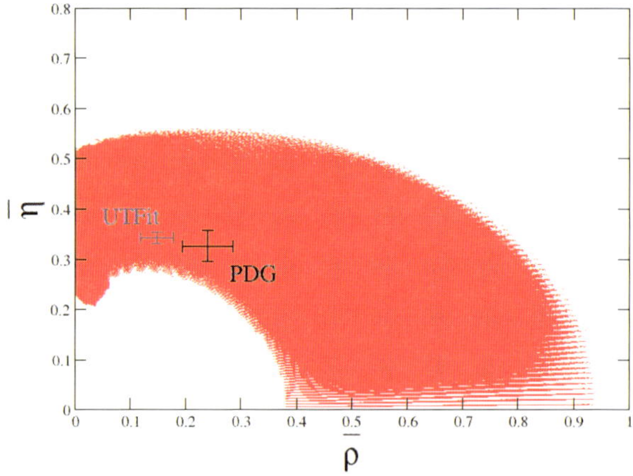

Fig. 2. The prediction of the model in the $\bar{\rho} - \bar{\eta}$ plane.[3]

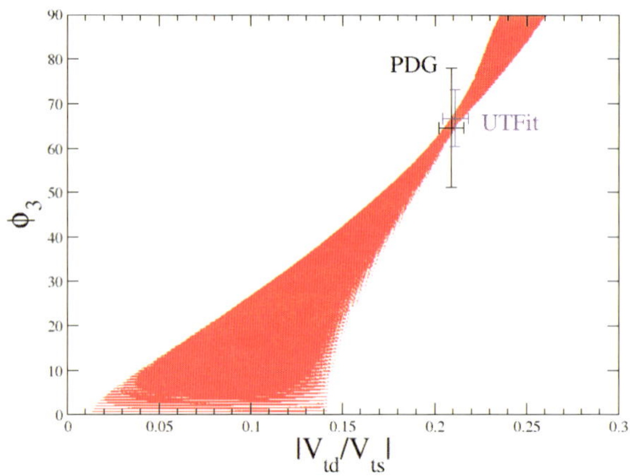

Fig. 3. The prediction of the model in the $|V_{td}/V_{ts}| - \phi_3(\gamma)$ plane.[3]

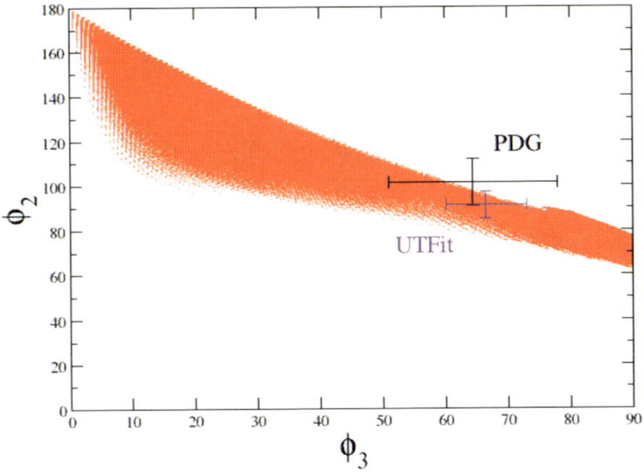

Fig. 4. The prediction of the model in the $\phi_3(\gamma)-\phi_2(\alpha)$ plane, where we used $m_d/m_b = (0.67 \sim 1.57) \times 10^{-3}$ and $m_s/m_b = (0.16 \sim 0.27) \times 10^{-1}$.[3]

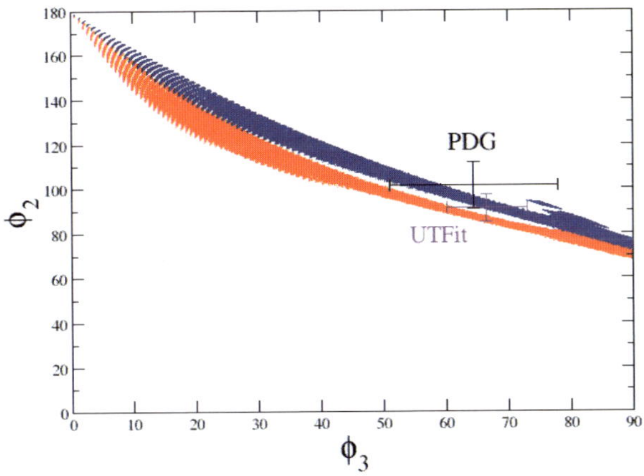

Fig. 5. The same as Fig. 4. Two predicted regions correspond to $m_s/m_b = 0.0215 \pm 0.00055$ and $m_d/m_b = (1.120 \pm 0.045)10^{-3}$ (red) and to $m_s/m_b = 0.0180 \pm 0.0005$ and $m_d/m_b = (0.900 \pm 0.045)10^{-3}$ (blue).

ON THE CONSEQUENCES OF TWISTED POINCARÉ SYMMETRY UPON QFT ON MOYAL NONCOMMUTATIVE SPACES

GAETANO FIORE*

Dipartimento di Matematica e Applicazioni, V. Claudio 21, 80125 Napoli, Italy
and
I.N.F.N., Sez. di Napoli, Complesso MSA, V. Cintia, 80126 Napoli, Italy
**E-mail: gfiore@na.infn.it*

We explore some general consequences of a consistent formulation of relativistic quantum field theory (QFT) on the Grönewold-Moyal-Weyl noncommutative versions of Minkowski space with covariance under the *twisted Poincaré group* of Chaichian *et al*,[12] Wess,[44] Koch *et al*,[31] Oeckl.[34] We argue that a proper enforcement of the latter requires *braided* commutation relations between any pair of coordinates $\hat{x}, \hat{y}$ generating two different copies of the space, or equivalently a $\star$*-tensor product* $f(x)\star g(y)$ (in the parlance of Aschieri *et al*[3]) between any two functions depending on x, y. Then all differences $(x-y)^\mu$ behave like their undeformed counterparts. Imposing (minimally adapted) Wightman axioms one finds that the n-point functions fulfill the same general properties as on commutative space. Actually, upon computation one finds (at least for scalar fields) that the n-point functions remain unchanged as functions of the coordinates' differences both if fields are free and if they interact (we treat interactions via time-ordered perturbation theory). The main, surprising outcome seems a QFT physically equivalent to the undeformed counterpart (to confirm it or not one should however first clarify the relation between n-point functions and observables, in particular S-matrix elements).

These results are mainly based on a joint work[24] with J. Wess.

Keywords: Quantum field theory; Noncommutative spaces; Moyal product; Quantum group symmetries

1. Introduction

The idea of spacetime noncommutativity is rather old. It goes back to Heisenberg[a]. The simplest noncommutativity one can think of is with co-

[a]Heisenberg proposed it in a letter to Peierls[29] to solve the problem of divergent integrals in relativistic quantum field theory. The idea propagated via Pauli to Oppenheimer. In

ordinates $\hat{x}^\mu$ fulfilling the commutation relations

$$[\hat{x}^\mu, \hat{x}^\nu] = i\theta^{\mu\nu}, \tag{1}$$

where $\theta^{\mu\nu}$ are the elements of a constant real antisymmetric matrix. Relations (1) have appeared in the literature under various names[b]. For brevity, we shall denote these noncommutative spaces as Moyal spaces. For present purposes $\mu = 0, 1, 2, 3$ and indices are raised or lowered through multiplication by the standard Minkowski metric $\eta_{\mu\nu}$, so as to obtain a deformation of Minkowski space. Clearly (1) are translation invariant, but not Lorentz-invariant (in 4 dimensions there is no isotropic antisymmetric 2-tensor $\theta^{\mu\nu}$). We shall denote by $\widehat{\mathcal{A}}$ the algebra"of functions on Moyal space", i.e. the algebra generated by $\mathbf{1}, \hat{x}^\mu$ fulfilling (1). For $\theta^{\mu\nu} = 0$ one obtains the algebra $\mathcal{A}$ generated by commuting x^μ.

Contributions to the construction of QFT on these spaces start in 1994-95.[17] A broad attention has been devoted to the program in the last decade, with a number of different approaches. By no means are they equivalent! Roughly speaking I would divide them into the following three groups.

Doplicher-Fredenhagen-Roberts (DFR) approach

This is field quantization in (rigorous) operator formalism on Moyal-Minkowski space, with usual Poincaré transformations. The pioneering works are,[17] the main developments can be found in.[5] Relations (1) are motivated by the interplay[c] of Quantum Mechanics and General Relativity in what Doplicher calls the **Principle of gravitational stability against localization of events:**

The gravitational field generated by the concentration of energy required by the Heisenberg Uncertainty Principle to localise an event in spacetime

---

1947 Snyder, a student of Oppenheimer, published the first proposal of a quantum theory built on a noncommutative space.[38]

[b]Sometimes they are called *canonical*, since by applying a Darboux transformation to the coordinates θ can be brought to canonical form (this depends only on its rank). More often the names contain some combination of the names of Weyl, Wigner, Grönewold, Moyal. This is due to the relation between canonical commutation relations and the $\star$-product (or twisted product) of Weyl and Von Neumann, which in turn was used by Wigner to introduce the Wigner transform; Wigner's work led Moyal to define the socalled Moyal bracket $[f \overset{\star}{,} g] = f \star g - g \star f$; the $\star$-product in position space [in the form of the asymptotic expansion of (10) with $x_i = x_j \equiv x$] first appeared in a paper by Grönewold.

[c]The arguments elaborate the well-known heuristic ones going back (as far as I know) to Wheeler.[45]

should not be so strong to hide[d] *the event itself to any distant observer - distant compared to the Planck scale.*[16,17]

In the first, simplest version $\theta^{\mu\nu}$ are not fixed constants, but central operators (obeying additional conditions) which on each irreducible representation become fixed constants $\sigma^{\mu\nu}$, the joint spectrum of $\theta^{\mu\nu}$. This allows to recover Lorentz covariance for the commutation relations. However, it seems that when developing the interacting theory the wished Lorentz covariance is sooner or later lost. In more recent versions $\theta^{\mu\nu}$ is no more central, but commutation relations remain of Lie-algebra type.

According to speculations heard in conference talks by Doplicher, $\theta^{\mu\nu}$ could be finally related to the vacuum expectation value (v.e.v.) of $R^{\mu\nu}$, which in turn should be influenced by the presence of matter quantum fields in spacetime (through quantum equations of motions).

Finally, we would like to mention the work,[27] which although not stricly in the DFR framework, also is based on a continuos family of fields labelled by the whole spectrum of noncommutative parameters $\theta^{\mu\nu}$, but has some overlap also with the following two approaches. A generalization of the procedure[27] has been proposed in the very recent work,[10] see also Buchholz's contribution to the present volume.

Path-integral quantization approach

This was initiated by Filk in[21] and has been adopted by most theoretical physicists, including many string-theorists, especially after the work.[37] Useful reviews are in.[18,41] The string-theorists' main motivation is that such models should describe the low-energy effective limit of string theory in a constant background B-field. Lorentz covariance [or $SO(4)$ covariance, after Wick-rotation] is lost, but this is expected in effective string theory because of the special direction selected by the B-field; only covariance under a subgroup[2] of $SL(2,\mathbb{C})$, the corresponding little group, is preserved.

The (Euclidean) classical field action used in the path-integral is deformed replacing products of fields by $\star$-products, whence modified Feynman rules for perturbative QFT are derived.

New complications seem to appear, like non-unitarity after naive Wick-rotation when $\theta^{0i} \neq 0$,[25] violation of causality,[9,36] mixing of UV and IR divergences[33] and subsequent non-renormalizability, alleged change of statistics, etc. Some of these problems, like non-unitarity or the very occurrence

[d] By black hole formation.

of divergences,[5] may be due simply to naive (and unjustified) applications of commutative QFT rules (path-integral methods, Feynman diagrams, analytic continuation, etc) and could disappear adopting the sounder field-operator approach. As for UV-IR mixing, while planar Feynman diagrams remain as the undeformed (apart from a phase factor), in particular have the same UV divergences, nonplanar Feynman diagrams which were UV divergent become finite for generic non-zero external momentum, but diverge as the latter go to zero, even with massive fields: these are the IR divergences. As a dramatic effect, infinitely many counterterms are necessary, making these theories non-renormalizable.

As a cure to the UV-IR mixing problem Grosse, Wulkenhaar[28] and collaborators add a x-dependent harmonic potential terms (e.g. $\sim \Omega^2 x^2 \varphi \star \varphi$ for a scalar field) to the Lagrangian (for a review see Grosse's contribution to the present volume, and references therein). Then the theory becomes renormalizable; actually $\Omega^2 x^2 \varphi \star \varphi$ is the only other marginal/relevant operator in the renormalization group flow. However these terms spoil the translation invariance of the theory.

Moreover, up to now no notion of Wick rotation between such QFT on Moyal-Euclidean space and QFT on Moyal-Minkowski noncommutative space has been found (there might be none).

Twisted Poincaré covariant approaches

These recover Poincaré covariance in a deformed version, following the observation[12,31,34,44] that (1) are *twisted Poincaré group* covariant. Field quantization is done either in a path-integral (on the Euclidean) or in an operator approach. The latter is the framework adopted in the present contribution; this is mainly based on the joint work[24] with J. Wess, who unfortunately has recently passed away.

How to implement twisted Poincaré covariance in QFT has been subject of debate and different proposals,[1,6–8,11,14,30,42,47] two main issues being whether one should: *a)* take the $\star$-product of fields at different spacetime points; *b)* deform the canonical commutation relations (CCR) of creation and annihilation operators $a, a^\dagger$ for free fields.

Our answers to questions *a)*, *b)* are affirmative and related to each other. The first arises from a proper analysis of twisted Poincaré transformations (section 2). In section 3 we adapt Wightman axioms to the noncommutative setting replacing all products by $\star$-products and analyze the consequences for Wightman and Green's functions. Motivated by the construction of normalizable states generated by the application to the vacuum of smeared fields (here we show why test functions in the Schwarz space are fine for smearing - a point we only briefly mentioned in[24]), we choose a setting where $\star$-products involve also the (Fock space) operator part of the fields; for free fields (section 4) this corresponds to choosing the second of the two options which were found admissible in[24] (they both lead to a $\star$-commutator of the fields equal to the undeformed counterpart). In section 4 we also briefly describe how the time-ordered perturbative computation of Green functions of a scalar $\varphi^{\star n}$-interacting theory gives the same results as the undeformed theory (the Feynman rules being unchanged). In section 5 we comment on what we can learn from these results, on which aspects still need investigation, and draw the conclusions.

2. Twisting Poincaré group and Minkowski spacetime

As already noted (1) are translation invariant, but not Lorentz-invariant. In[12,31,34,44] it has been recognized that they are however covariant under a deformed version of the Poincaré group, namely a triangular noncocommutative Hopf $*$-algebra H obtained from the Universal Enveloping algebra (UEA) $U\mathcal{P}$ of the Poincaré Lie algebra $\mathcal{P}$ by *twisting*[19e]. This means that (up to isomorphisms) H and $U\mathcal{P}$ (extended over the formal power series in $\theta^{\mu\nu}$) have

(1) the same $*$-algebra and counit ε (i.e. trivial representation);
(2) coproducts $\Delta, \hat{\Delta}$ related by

$$\Delta(g) \equiv \sum_I g^I_{(1)} \otimes g^I_{(2)} \quad \longrightarrow \quad \hat{\Delta}(g) = \mathcal{F}\Delta(g)\mathcal{F}^{-1} \equiv \sum_I g^I_{(\hat{1})} \otimes g^I_{(\hat{2})} \quad (2)$$

for any $g \in H \equiv U\mathcal{P}$. Fixed $\hat{\Delta}$, the socalled twist $\mathcal{F} \in H \otimes H$ is not uniquely determined, but what follows does not depend on its choice. The simplest is

$$\mathcal{F} \equiv \sum_I \mathcal{F}^{(1)}_I \otimes \mathcal{F}^{(2)}_I := \exp\left(\tfrac{i}{2}\theta^{\mu\nu} P_\mu \otimes P_\nu\right). \quad (3)$$

P_μ denote the generators of translations, and in (2), (3), we have used Sweedler notation; the $\sum_I$ may be a series, e.g. $\sum_I \mathcal{F}^{(1)}_I \otimes \mathcal{F}^{(2)}_I$ is the series arising from the power expansion of the exponential;

(3) antipodes $S, \hat{S}$ related by a similarity transformation; this is trivial for the above $\mathcal{F}$, so $\hat{S} = S$.

[e] In section 4.4.1 of[34] this was formulated in terms of the dual Hopf algebra.

For readers not familiar with Hopf algebras, we recall that the coproduct is the abstract operation by which one constructs the tensor product of any two representations. For the cocommutative Hopf algebra $U\mathbf{g}$ ($\mathbf{g}$ being a generic Lie algebra)

$$\Delta(\mathbf{1})=\mathbf{1}\otimes\mathbf{1}, \qquad g\in\mathbf{g}\rightarrow\Delta(g)=(g\otimes\mathbf{1}+\mathbf{1}\otimes g)\equiv g_1+g_2$$

and Δ is extended as a $*$-algebra map

$$\Delta: U\mathbf{g}\rightarrow U\mathbf{g}\otimes U\mathbf{g}, \qquad \Delta(ab)=\Delta(a)\Delta(b), \qquad \Delta(a^*)=[\Delta(a)]^{*\otimes *}. \tag{4}$$

The extension is unambiguous, as $\Delta\big([g,g']\big)=\big[\Delta(g),\Delta(g')\big]$ if $g,g'\in\mathbf{g}$. Also $\hat{\Delta}$ fulfills (4), $\hat{\Delta}(\mathbf{1})=\mathbf{1}\otimes\mathbf{1}$, as well as compatibility with ϵ and $*$ (as $\mathcal{F}$ is unitary). Then $\hat{\Delta}$ can replace Δ in constructing the tensor product of two representations of $U\mathbf{g}$. The antipode is the abstract operation by which one constructs the contragredient of any representation; it is uniquely determined by the coproduct, if it exists. In the present case, it is determined by $S(g)=-g$ if $g\in\mathbf{g}$, $S(\mathbf{1})=\mathbf{1}$, $S(ab)=S(b)S(a)$. Altogether, the structures $(U\mathcal{P},\cdot,*,\Delta,\epsilon,S)$, $(H,\cdot,*,\hat{\Delta},\epsilon,S)$ are examples of Hopf $*$-algebras (here we have explicitly indicated the algebra product by $\cdot$, but for brevity everywhere we shorten $a\cdot b=ab$).

For $U\mathcal{P}$ a straightforward computation gives

$$\hat{\Delta}(P_\mu)=P_\mu\otimes\mathbf{1}+\mathbf{1}\otimes P_\mu=\Delta(P_\mu), \quad \hat{\Delta}(M_\omega)=M_\omega\otimes\mathbf{1}+\mathbf{1}\otimes M_\omega+P[\omega,\theta]\otimes P\neq\Delta(M_\omega),$$

where we have set $M_\omega:=\omega^{\mu\nu}M_{\mu\nu}$ and used a row-by-column matrix product on the right. The left identity shows that the Hopf P-subalgebra remains undeformed and equivalent to the abelian translation group $\mathbb{R}^4$. Therefore, denoting by $\triangleright,\hat{\triangleright}$ the actions of $U\mathcal{P},H$ (on $\mathcal{A}$ $\triangleright$ amounts to the action of the corresponding algebra of differential operators, e.g. P_μ can be identified with $i\partial_\mu:=i\partial/\partial x^\mu$), they coincide on first degree polynomials a,b in $x^\nu,\hat{x}^\nu$,

$$P_\mu\triangleright x^\rho=i\delta^\rho_\mu=P_\mu\hat{\triangleright}\hat{x}^\rho, \qquad M_\omega\triangleright x^\rho=2i(x\omega)^\rho, \qquad M_\omega\hat{\triangleright}\hat{x}^\rho=2i(\hat{x}\omega)^\rho, \tag{5}$$

but $\triangleright,\hat{\triangleright}$ differ on higher degree polynomials in $x,\hat{x}$, as they are extended by the rules at the lhs of

$$g\triangleright(ab)=\textstyle\sum_I\big(g_{(1)}\triangleright a\big)\big(g_{(2)}\triangleright b\big) \tag{6}$$

$$g\,\hat{\triangleright}(\hat{a}\hat{b})=\textstyle\sum_I\big(g^I_{(\hat{1})}\hat{\triangleright}\hat{a}\big)\big(g^I_{(\hat{2})}\hat{\triangleright}\hat{b}\big) \;\Leftrightarrow\; g\triangleright(a\star b)=\sum_I\big(g^I_{(\hat{1})}\triangleright a\big)\star\big(g^I_{(\hat{2})}\triangleright b\big) \tag{7}$$

resp. involving the coproducts $\Delta(g),\hat{\Delta}(g)$ (these resp. reduce to the usual or a *deformed* Leibniz rule if $g=P_\mu,M_{\mu\nu}$). Moreover, $(g\triangleright a)^*=(Sg)^*\triangleright a^*$

as usual. Summarizing, the H-module unital $*$-algebra $\widehat{\mathcal{A}}$ is obtained by twisting the $U\mathcal{P}$-module unital $*$-algebra $\mathcal{A}$.

Several spacetime variables. Formulation through $\star$-products.

For $n \geq 1$ we denote the n-fold tensor product algebra of $\mathcal{A}$ by $\mathcal{A}^n$ and $x^\mu \otimes \mathbf{1} \otimes ..., \mathbf{1} \otimes x^\mu \otimes ...,...$ respectively by x_1^μ, x_2^μ, ... In other words, $\mathcal{A}^n$ is the algebra of functions of n sets of Minkowski coordinates x_i^μ, $i = 1, 2, ..., n$. The proper noncommutative deformation of $\mathcal{A}^n$ is the noncommutative unital $*$-algebra $\widehat{\mathcal{A}}^n$ generated by real variables $\hat{x}_i^\mu$ fulfilling the commutation relations at the lhs of

$$[\hat{x}_i^\mu, \hat{x}_j^\nu] = \mathbf{1} i\theta^{\mu\nu} \qquad \Leftrightarrow \qquad [x_i^\mu \stackrel{\star}{,} x_j^\nu] = \mathbf{1} i\theta^{\mu\nu}. \tag{8}$$

Note that the commutators are not zero for $i \neq j$; some authors erroneously impose (8) only for $i = j$. Relations (8) are compatible with the Leibniz rule $(7)_1$, so as to make $\widehat{\mathcal{A}}^n$ a H-module $*$-algebra, and are dictated by the *braiding* (see e.g.[32]) associated to the quasitriangular structure $\mathcal{R} = \mathcal{F}_{21}\mathcal{F}^{-1}$ of H; here $\mathcal{F}_{21} = \sum_I \mathcal{F}_I^{(2)} \otimes \mathcal{F}_I^{(1)}$.

As H is even triangular (i.e. $\mathcal{R}\mathcal{R}_{21} = \mathbf{1}^{\otimes 2}$), an essentially equivalent formulation of these H-module algebras is in terms of $\star$-products derived from $\mathcal{F}$. Denote by $\mathcal{A}_\theta^n$ the algebra obtained by endowing the vector space underlying $\mathcal{A}^n$ with a new product, the $\star$-product, related to the product in $\mathcal{A}^n$ by

$$a \star b := \textstyle\sum_I (\overline{\mathcal{F}}_I^{(1)} \triangleright a)(\overline{\mathcal{F}}_I^{(2)} \triangleright b), \tag{9}$$

with $\overline{\mathcal{F}} \equiv \mathcal{F}^{-1}$. This encodes both the usual $\star$-product within each copy of $\mathcal{A}$, and the "$\star$–tensor product" between different copies.[3,4] As a result one finds the isomorphic $\star$-commutation relations at the rhs of (8) [this follows from computing $x_i^\mu \star x_j^\nu$, which e.g. for the specific choice (3) gives $x_i^\mu x_j^\nu + i\theta^{\mu\nu}/2$] and that $\widehat{\mathcal{A}}^n, \mathcal{A}_\theta^n$ are isomorphic H-module unital $*$-algebras, in the sense of the equivalences (7), (8).

The $\star$-product (9) can be extended from polynomials a, b to power series. More explicitly, on analytic functions $a(x_i), b(x_j)$ (9) reads

$$a(x_i) \star b(x_j) = \exp[\frac{i}{2} \partial_{x_i} \theta \partial_{x_j}] a(x_i) b(x_j) \tag{10}$$

(for any 4-vectors p, q we define $p\theta q := p_\mu \theta^{\mu\nu} q_\nu$), what must be followed by the indentification $x_i = x_j$ *after* the action of the bi-pseudodifferential operator $\exp[\frac{i}{2}\partial_{x_i}\theta\partial_{x_j}]$ if $i = j$. Strictly speaking, the last formula makes sense if a, b belong to some suitable subalgebra[20] $\mathcal{A}^{n\prime}$ of the algebra of analytic functions such that the θ-power series is not only termwise well-defined but also convergent. Clearly $\mathcal{A}^{n\prime}$ will not be large enough for quantum-field-theoretic purposes. On the other hand, if $a(x_i), b(x_j) \in \mathcal{A}^{n\prime}$ admit Fourier transforms $\hat{a}(k_i), \hat{b}(k_j)$ then

$$a(x_i)\star b(x_j) = \int d^4k \int d^4q\, \hat{a}(k)\hat{b}(q) \exp[i(k\cdot x_i + q\cdot x_j - k\theta q/2)]. \tag{11}$$

This can be used as a *definition* of a $\star$-product for $a(x), b(x) \in L^1(\mathbb{R}^4) \cap \widehat{L^1(\mathbb{R}^4)}$, for $a(x) \in \mathcal{S}(\mathbb{R}^4)$ (Schwarz space) and $b(x) \in \mathcal{S}'(\mathbb{R}^4)$ (the space of tempered distributions), or conversely, as well as for $a(x), b(x) \in \mathcal{S}'(\mathbb{R}^4)$ provided $i \neq j$. These are in fact enough to reproduce all the product operations used in ordinary QFT, with results reducing to the commutative ones for $\theta^{\mu\nu}=0$.

Actually, for $i=j$ and some $a(x), b(x) \in \mathcal{S}'(\mathbb{R}^4)$ it may even happen that (11) is ill-defined for $\theta^{\mu\nu}=0$, but well-defined[26] (and thus "regularized") for $\theta^{\mu\nu}\neq 0$[f].

$\mathcal{S}(\mathbb{R}^4)$ is a $*$-module of the $*$-algebra underlying both $U\mathcal{P}, H$. As usual, the irreducible submodules are the eigenspaces of the Casimir $p{\cdot}p$; one can endow those characterized by a positive eigenvalue m^2 and a positive spectrum for P^0 by the usual pre-Hilbert space structure. By completion, one obtains unitary irreducible representations (irreps) of the $*$-algebra underlying both $U\mathcal{P}, H$, that describe scalar particles. (Generalized) eigenfunctions of P_μ or $M_{\mu\nu}$ exist instead within $\mathcal{S}'(\mathbb{R}^4)$, which is a larger $*$-module of the $*$-algebra underlying both $U\mathcal{P}, H$. Unitary irreps describing higher spin particles can be obtained in the standard way as some $\mathbb{C}^k \otimes_{\mathbb{C}} \mathcal{S}(\mathbb{R}^4)$ or projective modules thereof (spinor bundles, 4-vector bundles, etc). Summarizing, one obtains the same[12] classification (*à la* Wigner) of elementary particles as unitary irreps of either $U\mathcal{P}$ or H.

The generalization of the definition (11) to functions/distributions depending nontrivially on several (possibly all the) x_i is straightforward. In particular the $\star$-product $a \star b$ is well-defined for any $a \in \mathcal{S}(\mathbb{R}^{4n})$ and $b \in \mathcal{S}'(\mathbb{R}^{4n})$ (or viceversa). Also $\mathcal{S}(\mathbb{R}^{4n})$, $\mathcal{S}'(\mathbb{R}^{4n})$ are $*$-modules of the $*$-algebra underlying both $U\mathcal{P}, H$. In fact, we shall need to embed them in an even larger module $*$-algebra Φ^e of operator-valued (instead of c-number

[f]For instance, for $a(x) = \delta^4(x) = b(x)$ and invertible θ one easily finds $a(x_i)\star b(x_j) = (\pi^4 \det\theta)^{-1} \exp[2ix_j\theta^{-1}x_i]$; in particular for $i=j$ the exponential becomes 1 by the antisymmetry of θ^{-1}, and one finds a diverging constant as $\det\theta \to 0$, cf.[20,26] In[26] the largest algebra of distributions for which the $\star$-product is well-defined and associative is determined. In[20] the subalgebra of analytic functions for which (10) gives an asymptotic expansion of (11) is determined.

valued) distributions. The action $\triangleright$ fulfills the ordinary (resp. deformed) Leibniz rule (6) [resp. $(7)_2$] if a, b are multiplied (resp. $\star$-multiplied). This implies that the action of $U\mathcal{P}, H$ on tensor products of modules is constructed using the ordinary (resp. deformed) coproduct.

In the sequel we shall formulate the noncommutative spacetime only in terms of $\star$-products and construct QFT on it replacing all products by $\star$-products.

The **differential calculus is not deformed**, as $P_\mu \triangleright \partial_{x_i^\nu} = 0$ implies $\partial_{x_i^\nu}\star = \partial_{x_i^\nu} = \star\partial_{x_i^\nu}$:

$$\partial_{x_i^\mu} \star x_j^\nu = \delta_\mu^\nu \delta_j^i + x_j^\nu \star \partial_{x_i^\mu} \qquad\qquad \left[\partial_{x_i^\mu} \stackrel{\star}{,} \partial_{x_j^\nu}\right] = 0$$

($\hat{\partial}_{x_i^\mu}$ on $\widehat{\mathcal{A}}^n$ is isomorphic). In the sequel we shall drop the symbol $\star$ beside a derivative, as it has no effect. Also **integration over the space is not deformed**:

$$\int d^4x \, a(x) \star b(x) = \int d^4x \, a(x) b(x) \tag{12}$$

[this holds in particular for all $a(x) \in \mathcal{S}(\mathbb{R}^4)$ and $b(x) \in \mathcal{S}'(\mathbb{R}^4)$]. Stoke's theorem still applies. Using (11) it is easy to check the property

$$\int dx_i^4 \, b \star a(x_i) = b \star \int dx_i^4 \, a(x_i), \qquad \text{if } b \text{ is independent of } x_i, \tag{13}$$

analogous to the commutative conterpart [of course, if $a(x_i)$ is a c-number valued function/distribution depending only on x_i, the integral at the rhs is a c-number and the $\star$-product at the rhs can be dropped]. Therefore, for our purposes we can consider integration over any set of coordinates x as an operation commuting with the $\star$-product.

Let $a_i \in \mathbb{R}$ with $\sum_i a_i = 1$. An **alternative set of real generators of** $\mathcal{A}_\theta^n$ is:

$$\xi_i^\mu := x_i^\mu - x_{i+1}^\mu, \quad i = 1, \ldots, n-1, \qquad X^\mu := \textstyle\sum_{i=1}^n a_i x_i^\mu. \tag{14}$$

All ξ_i^μ are translation invariant, X^μ is not. It is immediate to check that $[X^\mu \stackrel{\star}{,} X^\nu] = \mathbf{1} i\theta^{\mu\nu}$, so X^μ generate a copy $\mathcal{A}_{\theta,X}$ of $\mathcal{A}_\theta$, whereas $\forall b \in \mathcal{A}_\theta^n$

$$\xi_i^\mu \star b = \xi_i^\mu b = b \star \xi_i^\mu \qquad \Rightarrow \qquad [\xi_i^\mu \stackrel{\star}{,} b] = 0, \tag{15}$$

so ξ_i^μ generate a $\star$-central subalgebra $\mathcal{A}_\xi^{n-1}$, and $\mathcal{A}_\theta^n \sim \mathcal{A}_\xi^{n-1} \otimes \mathcal{A}_{\theta,X}$. The $\star$-multiplication operators $\xi_i^\mu \star$ have the same spectral decomposition on all $\mathbb{R}$ (including 0) as multiplication operators $\xi^\mu \cdot$ by classical coordinates; the joint eigenvalues make up a space-like, or a null, or a time-like 4-vector, in the usual sense. Moreover, $\mathcal{A}_\xi^{n-1}, \mathcal{A}_{\theta,X}$ are actually H-module subalgebras, with

$$g \triangleright (a \star b) = \sum_I \left(g^I_{(1)} \triangleright a\right) \star \left(g^I_{(2)} \triangleright b\right), \qquad a \in \mathcal{A}_\xi^{n-1}, \quad b \in \mathcal{A}_\theta^n, \quad g \in H, \tag{16}$$

i.e. *on* $\mathcal{A}_\xi^{n-1}$ *the* H*-action is undeformed*, including the related part of the Leibniz rule. [By (15) here $\star$ can be also dropped].

Inverting (14), any set x_i can be expressed as a combination of the $n-1$ sets of $\star$-commutative variables ξ_i and the set X of $\star$-noncommutative ones, e.g. if $X := x_n$ then

$$x_i = \sum_{j=i}^{n-1} \xi_j + X.$$

X therefore behaves as parametrizing a "global noncommutative translation".

3. Revisiting Wightman axioms for QFT and their consequences

As in Ref.[40] we divide the Wightman axioms[39] into a subset (labelled by **QM**) encoding the quantum mechanical interpretation of the theory, its symmetry under space-time translations and stability, and a subset (labelled by **R**) encoding the relativistic properties. Since they provide minimal, basic requirements for the field-operator framework to quantization we try to apply them to the above noncommutative space (i.e. replacing everywhere products by $\star$-products) keeping the QM conditions, twisting Poincaré-covariance R1 and being ready to weaken locality R2 if necessary.

QM1. The states are described by vectors of a (separable) Hilbert space $\mathcal{H}$.

QM2. The group of space-time translations $\mathbb{R}^4$ is represented on $\mathcal{H}$ by strongly continuous unitary operators $U(a)$: the fields transform according to (26) with unit $A, U(A), \Lambda(A)$. The spectrum of the generators P_μ is contained in $\overline{V}_+ = \{p_\mu : p^2 \geq 0, p_0 \geq 0\}$. There is a unique Poincaré invariant state Ψ_0, the *vacuum state*.

QM3. The fields (in the Heisenberg representation) $\varphi^\alpha(x)$ [α enumerates field species and/or $SL(2,\mathbb{C})$-tensor components] are operator (on $\mathcal{H}$) valued tempered distributions on Minkowski space, with Ψ_0 a *cyclic* vector for the fields, i.e. $\star$-polynomials of the smeared fields applied to Ψ_0 give a set $\mathcal{D}_0$ dense in $\mathcal{H}$.

For a single scalar field $\mathcal{D}_0$ is the set of vectors of the form of a finite sum

$$\Psi_{\underline{f}} = f_0\Psi_0 + \varphi(f_1)\,\Psi_0 + \varphi\left(f_2^{(1)}\right)\star\varphi\left(f_2^{(2)}\right)\Psi_0 + ..., \tag{17}$$

where $f_j^{(h)} \in \mathcal{S}(\mathbb{R}^4)$, $h \leq j \leq N < \infty$ and

$$\varphi(f) := \int d^4x\, f(x)\star\varphi(x) \stackrel{(12)}{=} \int d^4x\, f(x)\varphi(x).$$

The (non-smeared) polynomials in the fields on commutative space make up a subalgebra Φ of what we may call the *(extended) field algebra* $\Phi^e = \left(\bigotimes_{i=1}^{\infty}\mathcal{S}'\right)\otimes\mathcal{O}$, where the first, second,... tensor factor $\mathcal{S}'$ is understood as the space of distributions depending on $\mathbf{x}_1, \mathbf{x}_2, ...$ [the dependence on $\mathbf{x}_h$ of the polynomial appearing in (17) being trivial for $h > N$], and $\mathcal{O}$ is the $*$-algebra of linear operators on $\mathcal{H}$ (e.g. for free bosonic/fermionic fields $\mathcal{O}$ is a Heisenberg/Clifford algebra with infinitely many modes). Φ^e also is a $U\mathcal{P}$-module $*$-algebra. We should therefore H-covariantly $\star$-deform the whole Φ^e into the corresponding Φ^e_θ (see also[23]). In analogy with the commutative case, we shall require that within Φ^e_θ fields $\star$-commute with c-number valued functions/distributions f

$$[\,\varphi^\alpha(x) \stackrel{\star}{,} f(y)\,] \equiv \varphi^\alpha(x)\star f(y) - f(y)\star\varphi^\alpha(x) = 0. \tag{18}$$

For free (scalar) fields this was proposed in[24] as the second of two admissible options (we shall explicitly recall how this works in section 3); this relation, together with (13), implies

$$\begin{aligned}
&\Psi_{\underline{f}} = f_0\Psi_0 + \!\int\! d^4x_1 f_1(x_1)\star\varphi(x_1)\Psi_0 + \!\int\! d^4x_1 \!\int\! d^4x_2 f_2(x_1,x_2)\star\varphi(x_1)\star\varphi(x_2)\Psi_0 + ...,\\
&f_j(x_1,...,x_j) := f_j^{(1)}(x_1)\star...\star f_j^{(j)}(x_j),
\end{aligned} \tag{19}$$

so $\Psi_{\underline{f}}$ is characterized by the terminating sequence $\underline{f} = (f_0, f_1, ...f_N)$. It is immediate to check that the Fourier transform of f_j differs from the commutative one only by a phase factor,

$$\tilde{f}_j(p_1,...,p_j) = \tilde{f}_j^{(1)}(p_1)...\tilde{f}_j^{(j)}(p_j)\exp\left[\frac{i}{2}\sum_{h=1}^{j}\sum_{k=h+1}^{j} p_h\theta p_k\right],$$

and therefore $f_j \in \mathcal{S}(\mathbb{R}^{4j})$. As on commutative space, $\mathcal{D}_0$ is also dense in the set $\mathcal{D}_1$ of all vectors of the form (19) with $f_j \in \mathcal{S}(\mathbb{R}^{4j})$.

Taking v.e.v.'s we define the *Wightman functions*

$$\mathcal{W}^{\alpha_1,\dots,\alpha_n}(x_1,\dots,x_n) := (\Psi_0, \varphi^{\alpha_1}(x_1)\star\dots\star\varphi^{\alpha_n}(x_n)\Psi_0), \qquad (20)$$

which are in fact distributions, and (their combinations) the *Green's functions*

$$G^{\alpha_1,\dots,\alpha_n}(x_1,\dots,x_n) := (\Psi_0, T[\varphi^{\alpha_1}(x_1)\star\dots\star\varphi^{\alpha_n}(x_n)]\Psi_0) \qquad (21)$$

where also *time-ordering* T is defined as on commutative space (even if $\theta^{0i}\neq 0$), e.g.

$$T[\varphi^{\alpha_1}(x)\star\varphi^{\alpha_2}(y)] = \varphi^{\alpha_1}(x)\star\varphi^{\alpha_2}(y)\star\vartheta(x^0-y^0) + \varphi^{\alpha_2}(y)\star\varphi^{\alpha_1}(x)\star\vartheta(y^0-x^0)$$

for $n=2$ (ϑ denotes the Heavyside function). This is well-defined as $\vartheta(x^0-y^0)$ is $\star$-central: the $\star$-products preceding all ϑ could be dropped, by (15).

Arguing as for ordinary QFT (see[39]) one finds that QM1-3 (alone) imply exactly the same properties as on commutative space:

W1. Wightman and Green's functions are translation-invariant tempered distributions and therefore may *depend only on the* ξ_i^μ:

$$\begin{aligned}\mathcal{W}^{\alpha_1,\dots,\alpha_n}(x_1,\dots,x_n) &= W^{\alpha_1,\dots,\alpha_n}(\xi_1,\dots,\xi_{n-1}),\\ \mathcal{G}^{\alpha_1,\dots,\alpha_n}(x_1,\dots,x_n) &= G^{\alpha_1,\dots,\alpha_n}(\xi_1,\dots,\xi_{n-1}).\end{aligned} \qquad (22)$$

W2. (**Spectral condition**) The support of the Fourier transform $\widetilde{W}$ of W is contained in the product of forward cones, i.e.

$$\widetilde{W}^{\{\alpha\}}(q_1,\dots q_{n-1}) = 0, \qquad \text{if } \exists j: \quad q_j \notin \overline{V}_+. \qquad (23)$$

From (19), (20) it follows that the scalar product of vectors $\Psi_{g_j} = \varphi\left(g_j^{(1)}\right)\star\dots\star\varphi\left(g_j^{(j)}\right)\Psi_0$, $\Psi_{f_k} = \varphi\left(f_k^{(1)}\right)\star\dots\star\varphi\left(f_k^{(k)}\right)\Psi_0$ is given by

$$(\Psi_{g_j},\Psi_{f_k}) = \int d^{4j}x\int d^{4k}y\, g_j^*(x_j,\dots,x_1)\star f_k(x_1,\dots,x_k)\star\mathcal{W}(x_1,\dots,x_j,y_1,\dots,y_k)$$

with f_k, g_j defined as in (19). Using (22) it is straightforward to prove[g] that in fact the previous formula holds also without $\star$ (as on commutative space):

$$(\Psi_{g_j},\Psi_{f_k}) = \int d^{4j}x\int d^{4k}y\, g_j^*(x_j,\dots,x_1) f_k(x_1,\dots,x_k)\mathcal{W}(x_1,\dots,x_j,y_1,\dots,y_k)\,. \qquad (24)$$

[g]The $\star$ between $\mathcal{W}$ and the rest is ineffective by $(15)_1$, $(20)_1$. Also the $\star$ between g_j^* and f_k is ineffective: going to the Fourier transforms, the corresponding phase factor reduces to 1 when exploiting the presence of the Dirac's δ in the equality $\widetilde{\mathcal{W}}(p_1,\dots,p_n) = (2\pi)^4\delta^4(\sum_i p_i)\widetilde{W}(p_1, p_1+p_2,\dots,p_1+\dots+p_n)$.

Using (24) (and the analogous formulae for non-scalar fields) we find

W3. $\mathcal{W}^{\{\alpha\}}$ fulfill the same **Hermiticity and Positivity** properties following from those of the scalar product in $\mathcal{H}$ as in the theory on commutative space.

For instance, for the Wightman functions of a single scalar field they reads as follows: $[\overline{\mathcal{W}(x_1,...,x_n)}]^* = \mathcal{W}(x_n,...,x_1)$, and for all terminating sequences $\underline{f} = (f_0, f_1, ...f_N)$ with $f_j \in \mathcal{S}(\mathbb{R}^{4j})$

$$(\Psi_{\underline{f}}, \Psi_{\underline{f}}) \equiv \sum_{j,k=1}^{\infty} \int d^{4j}x \int d^{4k}y\, f_j^*(x_j, ...x_1) f_k(y_1, ...y_k)\, \mathcal{W}(x_1, ...x_j, y_1, ...y_k) \geq 0. \tag{25}$$

The *ordinary* relativistic conditions on QFT are:

R1. (**Lorentz Covariance**) $SL(2,\mathbb{C})$ is represented on $\mathcal{H}$ by strongly continuous unitary operators $U(A)$, and under the Poincaré transformations $U(a,\, A) = U(a)\, U(A)$

$$U(a,A)\, \varphi^{\alpha}(x)\, U(a,A)^{-1} = S^{\alpha}_{\beta}(A^{-1})\, \varphi^{\beta}\big(\Lambda(A)x + a\big), \tag{26}$$

with S a finite-dimensional representation of $SL(2,\mathbb{C})$.

R2. (**Microcausality or locality**) The fields either commute or anticommute at spacelike separated points

$$[\,\varphi^{\alpha}(x), \varphi^{\beta}(y)\,]_{\mp} = 0, \qquad \text{for } (x-y)^2 < 0. \tag{27}$$

In *ordinary* QFT as a consequence of QM2,R1 one finds

W4. (**Lorentz Covariance of Wightman functions**)

$$\mathcal{W}^{\alpha_1..\alpha_n}\big(\Lambda(A)x_1, ..., \Lambda(A)x_n\big) = S^{\alpha_1}_{\beta_1}(A)..S^{\alpha_n}_{\beta_n}(A)\mathcal{W}^{\beta_1..\beta_n}(x_1, ..., x_n). \tag{28}$$

In particular, Wightman (and Green) functions of scalar fields are Lorentz invariant.

R1 needs a "twisted" reformulation **R1**$_\star$, which we defer. Now, however R1$_\star$ will look like, it should imply that $W^{\{\alpha\}}$ are $SL_\theta(2,\mathbb{C})$ tensors (in particular invariant if all involved fields are scalar). But, as the $W^{\{\alpha\}}$ are to be built only in terms of ξ_i^{μ} and other $SL(2,\mathbb{C})$ tensors (like $\partial_{x_i^\mu}$, $\eta_{\mu\nu}$, γ^{μ}, etc.), which are all annihilated by $P_\mu \triangleright$, $\mathcal{F}$ will act as the identity and $W^{\{\alpha\}}$ will transform under $SL(2,\mathbb{C})$ as for $\theta = 0$. Therefore **we shall require W4 also if** $\theta \neq 0$ as a temporary substitute of R1$_\star$.

The simplest sensible way to formulate the $\star$-analog of locality is

R2$_\star$. (Microcausality or locality) The fields either $\star$-commute or $\star$-anticommute at spacelike separated points

$$[\,\varphi^\alpha(x) \stackrel{\star}{,} \varphi^\beta(y)\,]_\mp = 0, \qquad \text{for } (x-y)^2 < 0. \tag{29}$$

This makes sense, as space-like separation is sharply defined, and reduces to the usual locality when $\theta = 0$. Therefore we shall adopt it. Whether there exist reasonable weakenings of R2$_\star$ is an open question also on commutative space, and the same restrictions will apply.

Arguing as in[39] one proves that QM1-3, W4, R2$_\star$ are independent and compatible, as they are fulfilled by free fields (see below): the noncommutativity of a Moyal-Minkowski space is compatible with R2$_\star$! As consequences of R2$_\star$ one again finds

W5. (Locality) if $(x_j - x_{j+1})^2 < 0$

$$\mathcal{W}(x_1, ...x_j, x_{j+1}, ...x_n) = \pm\mathcal{W}(x_1, ...x_{j+1}, x_j, ...x_n). \tag{30}$$

W6. (Cluster property) For any spacelike a and for $\lambda \to \infty$

$$\mathcal{W}(x_1, ...x_j, x_{j+1} + \lambda a, ..., x_n + \lambda a) \to \mathcal{W}(x_1, ..., x_j)\,\mathcal{W}(x_{j+1}, ..., x_n), \tag{31}$$

(convergence in the distribution sense); this is true also with permuted x_i's.

Summarizing: our QFT framework is based on **QM1-3, W4, R2$_\star$** and the technical requirement (18), or alternatively on the constraints **W1-6** for $\mathcal{W}^{\{\alpha\}}$, exactly as in QFT on Minkowski space. We stress that this applies for all $\theta^{\mu\nu}$, even if $\theta^{0i} \neq 0$, contrary to other approaches. Moreover, we have just seen that (contrary to[13]) we can keep the Schwarz space $\mathcal{S}(\mathbb{R}^4)$ as the space of test functions for smearing the fields. We shall keep it as this guarantees not only the separability of $\mathcal{H}$ but also that a finite number of subtractions is enough to define field products at the same point, i.e. essentially the possibility to renormalize the theory. However we should note that, for given $f_j^{(h)} \in \mathcal{S}(\mathbb{R}^4)$, the states (17) do not coincide with their undeformed counterparts. We do not know whether this might have consequences on observables (as S-matrix elements).

4. Free or interacting scalar field

As the differential calculus remains undeformed, so remain the equation of motions of free fields. Sticking for simplicity to the case of a scalar field of mass m, the solution of the Klein-Gordon equation reads

$$\varphi_0(x) = \int d\mu(p)\,[e^{-ip\cdot x} \star a^p + a_p^\dagger \star e^{ip\cdot x}\,] \tag{32}$$

where $d\mu(p) = \delta(p^2 - m^2)\vartheta(p^0)d^4p = dp^0\delta(p^0 - \omega_{\mathbf{p}})d^3\mathbf{p}/2\omega_{\mathbf{p}}$ is the invariant measure ($\omega_{\mathbf{p}} := \sqrt{\mathbf{p}^2 + m^2}$). Postulating the axioms of the preceding section, except $\mathbf{R2}_\star$, one can prove that up to a positive factor (which can be always reabsorbed in a field redefinition)

$$\begin{aligned} W(x-y) &= \int d\mu(p) e^{-ip\cdot(x-y)} \\ G(x-y) &= -i \int \frac{d^4p}{2\pi} \frac{e^{-ip\cdot(x-y)}}{p^2 - m^2 + i\epsilon}, \end{aligned} \tag{33}$$

and therefore coincides with the undeformed counterpart. Adding also $\mathbf{R2}_\star$ one can prove the **free field commutation relation**

$$[\varphi_0(x) \stackrel{\star}{,} \varphi_0(y)] = 2 \int d\mu(p) \, \sin[p\cdot(x-y)] =: iF(x-y), \tag{34}$$

coinciding with the undeformed one. Applying ∂_{y^0} to (34) and setting $y^0 = x^0$ [this is compatible with (8)] one finds **the canonical commutation relation**

$$[\varphi_0(x^0, \mathbf{x}) \stackrel{\star}{,} \dot{\varphi}_0(x^0, \mathbf{y})] = i\,\delta^3(\mathbf{x} - \mathbf{y}). \tag{35}$$

As a consequence of (34), **the n-point Wightman functions** not only fulfill W1-W6, but **coincide with the undeformed ones**, i.e. vanish if n is odd and are sum of products of 2-point functions (factorization) if n is even.

A φ_0 fulfilling (34) can be obtained assuming $P_\mu \triangleright a^\dagger_p = p_\mu a^\dagger_p$, $P_\mu \triangleright a^p = -p_\mu a^p$, so as to extend the $\star$-product law also to $a^p, a^\dagger_p$, and plugging in (32) $a^p, a^\dagger_p$ satisfying

$$\begin{aligned} &a^\dagger_p \star a^\dagger_q = e^{-ip\theta q}\, a^\dagger_q \star a^\dagger_p, \qquad a^p \star a^q = e^{-ip\theta q}\, a^q \star a^p, \\ &a^p \star a^\dagger_q = e^{ip\theta q}\, a^\dagger_q \star a^p + 2\omega_{\mathbf{p}}\delta^3(\mathbf{p} - \mathbf{q}), \\ &a^p \star e^{iq\cdot x} = e^{-ip\theta q}\, e^{iq\cdot x} \star a^p, \qquad a^\dagger_p \star e^{iq\cdot x} = e^{ip\theta q}\, e^{iq\cdot x} \star a^\dagger_p. \end{aligned} \tag{36}$$

Note the nontrivial commutation relations between the $a^p, a^\dagger_p$ and c-number valued functions, but $[\varphi_0(x) \stackrel{\star}{,} f(y)] = 0$ as in (18). The first three relations define an example of a general deformed Heisenberg algebra[22]

$$\begin{aligned} &a^q \star a^p = R^{qp}_{rs}\, a^s \star a^r, \qquad a^\dagger_p \star a^\dagger_q = R^{sr}_{pq}\, a^\dagger_r \star a^\dagger_s, \\ &a^p \star a^\dagger_q = \delta^p_q + R^{rp}_{qs}\, a^\dagger_r \star a^s, \end{aligned} \tag{37}$$

covariant under a triangular Hopf algebra H. Here $\mathcal{R} := \mathcal{F}_{21}\mathcal{F}^{-1}$ is the triangular structure of H, $\{|p\rangle\}$ is the generalized basis of the 1-particle Hilbert space consisting of (on-shell) eigenvectors of P_μ, $\delta^p_q = 2\omega_{\mathbf{p}}\delta^3(\mathbf{p} - \mathbf{q})$ is Dirac's delta (up to normalization), $R^{pq}_{rs} := \langle\mathbf{p}|\otimes\langle\mathbf{q}|\mathcal{R}\,|\mathbf{r}\rangle\otimes|\mathbf{s}\rangle = e^{ip\theta q}\delta^p_r\delta^q_s$.

Up to normalization of R, and with $p, q, r, s \in \{1, ..., N\}$, relations (25) are also identical to the ones defining the older q-deformed Heisenberg algebras of,[35,46] based on a quasitriangular $\mathcal{R}$ in (only) the N-*dimensional* representation of $H = U_q su(N)$.

Remark. In[24] we actually found also a different (and maybe more intuitive) way to construct a free field fulfilling (34). It amounts to: 1. introducing $a^p, a_p^\dagger$ satisfying

$$a_p^\dagger a_q^\dagger = e^{ip\theta' q}\, a_q^\dagger a_p^\dagger, \quad a^p a^q = e^{ip\theta' q}\, a^q a^p, \quad a^p a_q^\dagger = e^{-ip\theta' q}\, a_q^\dagger a^p + 2\omega_{\mathbf{p}} \delta^3(\mathbf{p}-\mathbf{q}),$$

$$(\text{with } \theta' = \theta), \qquad \text{and } [a^p, f(x)] = [a_p^\dagger, f(x)] = 0, \tag{38}$$

(so c-number valued functions/distributions keep commuting with $a^p, a_p^\dagger$), as adopted e.g. in;[1,7,30] 2. restricting $\star$-multiplication only to the functions/distributions part (i.e. elements of the extended $\mathcal{A}_\theta^n$) of the fields. Consequently, instead of (32) the field decomposition reads $\varphi_0(x) = \int d\mu(p)\, [e^{-ip\cdot x} a^p + a_p^\dagger e^{ip\cdot x}]$ with such $a^p, a_p^\dagger$. This leads to the same properties **W1-W6**. However, as $\varphi(f)$ does no more depend on spacetime coordinates x, the $\star$ in (17) and (19) becomes redundant, and we obtain $\Psi_{\underline{f}} = f_0 \Psi_0 + \int d^4x_1 f_1(x_1) \varphi(x_1)\, \Psi_0 + \int d^4x_1 \int d^4x_2 f_2(x_1, x_2) \varphi_0(x_1) \varphi_0(x_2)\, \Psi_0 + ...$, with $f_j(x_1, ..., x_j) = f_j^{(1)}(x_1) ... f_j^{(j)}(x_j)$. As a result, scalar products (Ψ_{g_j}, Ψ_{f_k}) cannot be expressed in terms of Wightman functions as in (24), but in the form

$$(\Psi_{g_j}, \Psi_{f_k}) = \int d^{4j}x \int d^{4k}y\, g_j^*(x_j, ..., x_1) f_k(x_1, ..., x_k) \mathcal{W}'(x_1, ..., x_j, y_1, ..., y_k)$$

$$\mathcal{W}'(x_1, ..., x_j, y_1, ..., y_k) := (\Psi_0, \varphi_0(x_1) ... \varphi_0(x_j) \varphi_0(y_1) ... \varphi_0(y_k) \Psi_0)$$

(with no $\star$-products in the definition of $\mathcal{W}'$, as in[8]]). The distributions $\mathcal{W}'$ do not fulfill all the properties W1-W6 (except of course in the undeformed case $\theta' = 0$). We also briefly consider some consequences of choosing $\theta' \neq \theta$ in (38) ($\theta' = 0$ gives CCR among the $a^p, a_p^\dagger$, assumed in most of the literature, explicitly[17] or implicitly, in operator[14,15] or in path-integral approach to quantization) together with $\varphi_0(x) = \int d\mu(p)\, [e^{-ip\cdot x} a^p + a_p^\dagger e^{ip\cdot x}]$ and definition $(20)_1$ for the Wightman functions. One finds the non-local $\star$-commutation relation

$$\varphi_0(x) \star \varphi_0(y) = e^{i\partial_x (\theta - \theta') \partial_y} \varphi_0(x) \star \varphi_0(y) + i\, F(x - y),$$

and the corresponding (free field) Wightman functions violate W4, W6, unless $\theta' = \theta$.

Going back to our framework, we now define **normal ordering** as a $\mathcal{A}_\theta^n$-bilinear map of field algebra into itself such that $(\Psi_0, :M: \Psi_0) = 0$ fo any field polynomial M, in particular $:\mathbf{1}: = 0$. Applying it to (38) we find that it is consistent to define

$$:a^p \star a^q := a^p \star a^q, \quad :a_p^\dagger \star a^q := a_p^\dagger \star a^q, \quad :a_p^\dagger \star a_q^\dagger := a_p^\dagger \star a_q^\dagger, \quad :a^p \star a_q^\dagger := a_q^\dagger \star a^p e^{-ip\theta q}$$

(note the phase). More generally, by definition in any monomial this map reorders all a^p to the right of all $a_q^\dagger$ introducing a $e^{-iq\theta p}$ for each flip $a^p \leftrightarrow a_q^\dagger$. For $\theta = 0$ the map reduces to the undeformed normal ordering.

As a result, one finds that the v.e.v. of any normal-ordered $\star$-polynomial of fields is zero, that normal-ordered $\star$-products of fields can be obtained from $\star$-products by the undeformed pattern of subtractions, and that **the same Wick theorem** as in the undeformed case holds. Applying **time-ordered perturbation theory** to an interacting field again one can heuristically derive,[24] through the same arguments used on commutative space, the Gell-Mann–Low formula

$$G(x_1, ..., x_n) = \frac{\left(\Psi_0, T\left\{\varphi_0(x_1) \star ... \star \varphi_0(x_n) \star \exp\left[-i\lambda \int dy^0 \; H_I(y^0)\right]\right\} \Psi_0\right)}{\left(\Psi_0, T \exp\left[-i \int dy^0 \; H_I(y^0)\right] \Psi_0\right)} \tag{39}$$

(which is rigorously valid under the assumption of asymptotic completeness, $\mathcal{H} = \mathcal{H}^{in} = \mathcal{H}^{out}$). Here $\varphi_0, H_I(x^0)$ denote the free "in" field (i.e. the incoming field) and the interaction Hamiltonian in the interaction representation, e.g.

$$H_I(x^0) = \lambda \int d^3x \; :\varphi_0^{\star m}(x): \star, \qquad \varphi_0^{\star m}(x) \equiv \underbrace{\varphi_0(x) \star ... \star \varphi_0(x)}_{m \text{ times}}. \tag{40}$$

Thus[24] one finds that the **Green functions (39) coincide with the undeformed ones** (at least perturbatively). They can be computed by Feynman diagrams with the undeformed Feynman rules, and the theory can be regularized and renormalized in the standard ways.

5. Conclusions. What do we learn?

Although various approaches to relativistic QFT on Moyal-Minkowski space have been proposed, there is still no generally accepted one. Operator-based approaches look safer starting points, but twisting or not the Poincaré group, and doing it properly, makes the results radically different.

We have claimed here that a sensible theory with twisted Poincaré seems possible and avoids all complications (IV-UR, causality/unitarity violation, statistics violation, cluster property violation, loss of spacetime symmetry,...). It naturally involves a compensation of operator $(a, a^\dagger)$ and spacetime noncommutativities, so that the free field $\star$-commutators coincide with the undeformed ones.

The surprising and probably disappointing fact is that also the corresponding n-point functions, expressed as functions of the coordinates' differences, coincide with the undeformed ones. The natural consequence seems that no new physics, nor a more satisfactory formulation of the old one (e.g. by an intrinsic UV regularization) is obtained (at least for scalar fields), although this can be confirmed only upon clarifying the relation between n-point functions and observables, in particular S-matrix elements.

Nevertheless we think that we can learn quite much from trying to understand the reasons of these surprising results, which are in striking contrast with the ones found in most of the literature, as well as from using our approach as a laboratory for:

(1) searching and testing equivalent formulations of QFT on NC spaces: Wick rotation into EQFT, path integral quantization, etc.;
(2) clarifying notions such as asymptotic states, spin-statistics, CPT, etc., on noncommutative spaces;
(3) properly formulating covariance properties of fields under twisted symmetries ($\mathrm{R1}_\star$), and clarify their connection to the ordinary ones;
(4) properly formulating gauge field theory on noncommutative spaces.

Acknowledgments

I would like to thank Profs. W. Zimmermann, E. Seiler and K. Sibold for the very kind invitation to the "Zimmermannfest 08" conference, and for the warm atmosphere experienced there.

References

1. Y. Abe, "Noncommutative Quantization for Noncommutative Field Theory", Int. J. Mod. Phys. **A22** (2007), 1181-1200.
2. L. Alvarez-Gaume, M. A. Vazquez-Mozo, "General Properties of Noncommutative Field Theories", Nucl. Phys. **B668** (2003), 293-321.
3. P. Aschieri, C. Blohmann, M. Dimitrijevic, F. Meyer, P. Schupp, J. Wess, "Noncommutative Geometry and Gravity", Class. Quant. Grav. **22** (2005), 3511-3532.
4. P. Aschieri, M. Dimitrijevic, F. Meyer, J. Wess, "Noncommutative Geometry and Gravity", Class. Quant. Grav. **23** (2006), 1883-1912.
5. D. Bahns, S. Doplicher, K. Fredenhagen, G. Piacitelli, "On the unitarity problem in space/time noncommutative theories", Phys. Lett. **B533** (2002),

178-181; "Ultraviolet Finite Quantum Field Theory on Quantum Spacetime", Commun. Math. Phys. **237** (2003), 221-241; "Field Theory on Noncommutative Spacetimes: Quasiplanar Wick Products", Phys. Rev. **D71** (2005), 025022.

6. A. P. Balachandran, G. Mangano, A. Pinzul, S. Vaidya, "Spin and Statistics on the Groenewold-Moyal Plane: Pauli-Forbidden Levels and Transitions", Int. J. Mod. Phys. **A21** (2006), 3111-3126.
7. A. P. Balachandran, A. Pinzul, B. A. Qureshi, "UV-IR Mixing in Non-Commutative Plane", Phys. Lett. **B634** (2006), 434-436.
8. A. P. Balachandran, T. R. Govindarajan, G. Mangano, A. Pinzul, B. A. Qureshi, S. Vaidya, "Statistics and UV-IR Mixing with Twisted Poincaré Invariance", Phys. Rev. **D75** (2007), 045009.
9. H. Bozkaya, P. Fischer, H. Grosse, M. Pitschmann, V. Putz, M. Schweda, R. Wulkenhaar, "Space/time noncommutative field theories and causality", Eur. Phys. J. **C29** (2003), 133-141.
10. D. Buchholz, S. J. Summers, "Warped Convolutions: A Novel Tool in the Construction of Quantum Field Theories", arXiv:0806.0349 and contribution to this volume.
11. J.-G. Bu, H.-C. Kim, Y. Lee, C. H. Vac, J. H. Yee, "Noncommutative Field Theory from twisted Fock space" Phys. Rev. **D73** (2006), 125001.
12. M. Chaichian, P. Kulish, K. Nishijima, A. Tureanu, "On a Lorentz-Invariant Interpretation of Noncommutative Space-Time and Its Implications on Noncommutative QFT", Phys. Lett. **B604** (2004), 98-102.
13. M. Chaichian, M. Mnatsakanova, A. Tureanu, Yu. Vernov, "Test Functions Space in Noncommutative Quantum Field Theory" arXiv:0706.1712.
14. M. Chaichian, P. Presnajder, A. Tureanu, "New concept of relativistic invariance in NC space-time: twisted Poincaré symmetry and its implications", Phys. Rev. Lett. **94** (2005), 151602.
15. M. Chaichian, M. Mnatsakanova, K. Nishijima, A. Tureanu, Yu. Vernov, "Towards an Axiomatic Formulation of Noncommutative Quantum Field Theory", hep-th/0402212.
16. S. Doplicher, "Quantum Field Theory on Quantum Spacetime", hep-th/0608124.
17. S. Doplicher, K. Fredenhagen, J. E. Roberts, "The quantum structure of spacetime at the Planck scale and quantum fields", Commun. Math. Phys. **172** (1995), 187-220; "Spacetime Quantization Induced by Classical Gravity", Phys. Lett. **B 331** (1994), 39-44.
18. M. R. Douglas, N. Nekrasov, "Noncommutative Field Theory", Rev. Mod. Phys. **73** (2001), 977-1029.
19. V. G. Drinfeld, "On constant quasiclassical solutions of the Yang-Baxter quantum equation", Sov. Math. Dokl. **28** (1983), 667.
20. R. Estrada, J. M. Gracia-Bondía, J. C. Várilly, "On asymptotic expansions of twisted products", J. Math. Phys. **30** (1989), 2789-2796.
21. T. Filk, "Divergencies in a field theory on quantum space", Phys. Lett. **B376** (1996), 53-58.
22. G. Fiore, "Deforming Maps for Lie Group Covariant Creation & Annihilation

Operators", J. Math. Phys. **39** (1998), 3437-3452.

23. G. Fiore, "On second quantization on noncommutative spaces with twisted symmetries", in preparation.
24. G. Fiore, J. Wess, "Full twisted Poincare' symmetry and quantum field theory on Moyal-Weyl spaces", Phys. Rev. **D75** (2007), 105022.
25. J. Gomis, T. Mehen, "Space-Time Noncommutative Field Theories And Unitarity", Nucl.Phys. **B591** (2000), 265-276.
26. J. M. Gracia-Bondía, J. C. Várilly, "Algebras of distributions suitable for phase-space quantum mechanics. I", J. Math. Phys. **29** (1988), 869-879.
27. H. Grosse, G. Lechner, "Wedge-Local Quantum Fields and Noncommutative Minkowski Space", JHEP **0711**:012, 2007.
28. H. Grosse and R. Wulkenhaar, "Renormalisation of phi**4 theory on noncommutative R**2 in the matrix base", **JHEP 12** (2003), 019; "Renormalisation of phi**4 theory on non-commutative R**4 in the matrix base", Commun. Math. Phys. **256** (2005), 305-374.
29. *Letter of Heisenberg to Peierls* (1930), in: Wolfgang Pauli, Scientific Correspondence, vol. II, 15, Ed. Karl von Meyenn, Springer-Verlag 1985.
30. F. Lizzi, S. Vaidya, P. Vitale, "Twisted Conformal Symmetry in Noncommutative Two-Dimensional Quantum Field Theory", Phys. Rev. **D73** (2006), 125020.
31. F. Koch, E. Tsouchnika, "Construction of θ-Poincaré Algebras and their Invariants on $\mathcal{M}_\theta$", Nucl.Phys. **B717** (2005), 387-403.
32. S. Majid, *Foundations of Quantum Groups*, Cambridge Univ. Press (1995); and references therein.
33. S. Minwalla, M. Van Raamsdonk, N. Seiberg, "Noncommutative pertur- bative dynamics", **JHEP 02** (2000) 020.
34. R. Oeckl, "Untwisting Noncommutative $\mathbb{R}^d$ and the Equivalence of Quantum Field Theory", Nucl. Phys. **B581** (2000), 559-574.
35. W. Pusz, S. L. Woronowicz, "Twisted Second Quantization", Reports on Mathematical Physics **27** (1989), 231.
36. N. Seiberg, L. Susskind, N. Toumbas, Space/Time Non-Commutativity and Causality, JHEP **0006** (2000), 044.
37. N. Seiberg, E. Witten, "String Theory and Noncommutative Geometry", JHEP **9909** (1999), 032.
38. H. S. Snyder, "Quantized spacetime", Phys.Rev. **71** (1947), 38.
39. R. F. Streater, A. S. Wightman, *PCT, Spin and Statistics and All That*, Benjamin 1964.
40. F. Strocchi, *Selected topics on the general properties of quantum field theory*, Lecture notes in physics 51, World Scientific, Singapore [etc.], 1993; "Relativistic Quantum Mechanics and Field Theory" Found. Phys. **34** (2004) 501-527.
41. R. J. Szabo, "Quantum Field Theory on Noncommutative Spaces", Phys. Rept. **378** (2003), 207-299.
42. A. Tureanu, "Twist and Spin-Statistics Relation in Noncommutative Quantum Field Theory", Phys. Lett. **B638** (2006), 296-301.
43. M. Van Raamsdonk and N. Seiberg, "Comments on noncommutative pertur-

bative dynamics", **JHEP 03** (2000), 035.
44. J. Wess, "Deformed Coordinate Spaces; Derivatives", Lecture given at BW2003 Workshop, hep-th/0408080
45. J. A. Wheeler, "Geometrodynamics and the Issue of the Final State". In: Relativity, Groups and Topology. De Witt, C., De Witt, B., (eds.) Gordon and Breach 1965, 317-520.
46. J. Wess and B. Zumino, "Covariant Differential Calculus On The Quantum Hyperplane", Nucl. Phys. Proc. Suppl. **18B** (1991), 302.
47. J. Zahn, "Remarks on twisted noncommutative quantum field theory", Phys. Rev. **D73** (2006), 105005.

TAMING THE LANDAU GHOST IN NONCOMMUTATIVE QUANTUM FIELD THEORY

HARALD GROSSE*

Physics Department, University of Vienna
A-1090 Vienna, Austria
**E-mail: harald.grosse@univie.ac.at*

After a short introduction into the formulation of noncommutative field theory and the discussion of the IR/UV mixing, I review the main ideas and techniques of our proof with Raimar Wulkenhaar that the duality-covariant four-dimensional noncommutative scalar model is renormalizable to all orders. Next I discuss the calculation of the one-loop contribution to the beta function and emphasize the taming of the Landau ghost.

I continue with the formulation of fermion models as well as gauge models, where less results are worked out.

For fields defined over a deformed Minkowski space-time the property of Wedge-locality replacing locality is mentioned.

Keywords: Noncommutative geometry; Quantum field theory; Landau ghost; Renormalization; Wedge-locality

1. Preface

Four years ago, I obtained a phone call from Prof. Zimmermann asking me about the status of renormalizability of noncommutative quantum field theory. At that time, we did not have finished our proof, but of course I enjoyed very much the interest of Prof. Zimmermann. In addition I remember several times discussions with Prof. Zimmermann when I was visiting the Max-Planck Institute at Föhringer Ring. It encouraged us to continue our work on this subject.

I dedicate this review to Prof. Zimmermann and I wish you many happy recurrences.

2. Introduction

Quantum field theory on Euclidean or Minkowski space is extremely successful. For suitably chosen action functionals one achieves a remarkable agreement of up to 10^{-11} between theoretical predictions and experimental data. However, combining the fundamental principles of both general relativity and quantum mechanics one concludes that space(-time) cannot be a differentiable manifold[1,2]

To make this transparent, let us ask how we explore technically the geometry of space(-time). The building blocks of a manifold are the points labelled by coordinates $\{x^\mu\}$ in a given chart. Points enter quantum field theory via the values of the fields at the point labelled by $\{x^\mu\}$. This observation provides a way to "visualise" the points: we have to prepare a distribution of matter which is sharply localised around $\{x^\mu\}$. For a perfect visualisation we need a δ-distribution of the matter field. This is physically not possible, but one would think that a δ-distribution could be arbitrarily well approximated. However, that is not the case, there are limits of localisability long before the δ-distribution is reached.

Let us assume that there is a matter distribution which is believed to have two separated peaks within a space-time region R of diameter d. How do we test this conjecture? We perform a scattering experiment in the hope to find interferences which tells us about the internal structure in the region R. We clearly need test particles of de Broglie wave length $\lambda = \frac{\hbar c}{E} \lesssim \mathrm{d}$, otherwise we can only resolve a single peak. For $\lambda \to 0$ the gravitational field of the test particles becomes important. The gravitational field created by an energy E can be measured in terms of the Schwarzschild radius

$$r_s = \frac{2G_N E}{c^4} = \frac{2G_N \hbar}{\lambda c^3} \gtrsim \frac{2G_N \hbar}{\mathrm{d}c^3} , \tag{1}$$

where G_N is Newton's constant. If the Schwarzschild radius r_s becomes larger than the radius $\frac{\mathrm{d}}{2}$, the inner structure of the region R can no longer be resolved (it is behind the horizon). Thus, $\frac{\mathrm{d}}{2} \geq r_s$ leads to the condition

$$\frac{\mathrm{d}}{2} \gtrsim \ell_P := \sqrt{\frac{G_N \hbar}{c^3}} , \tag{2}$$

which means that the Planck length ℓ_P is the fundamental length scale below which length measurements become meaningless. Space-time cannot be a manifold.

Since geometric concepts are indispensable in physics, we need a replacement for the space-time manifold which still has a geometric interpretation. Quantum physics tells us that whenever there are measurement limits we have to describe the situation by non-commuting operators on a Hilbert space. Fortunately for physics, mathematicians have developed a generalization of geometry, called noncommutative geometry,[3] which is perfectly designed for our purpose. However, in physics we need more than just a better geometry: We need renormalisable quantum field theories modelled on such a noncommutative geometry.

Remarkably, it turned out to be very difficult to renormalise quantum field theories even on the simplest noncommutative spaces.[4] It would be a wrong conclusion, however, that this problem singles out the standard commutative geometry as the only one compatible with quantum field theory. The problem tells us that we are still at the very beginning of *understanding* quantum field theory. In doing quantum field theory on noncommutative geometries we learn a lot about quantum field theory itself.

3. Formulation of nc QFT on Moyal space

The simplest noncommutative generalisation of Euclidean space is the so-called noncommutative $\mathbb{R}^D$. Although this space arises naturally in a certain limit of string theory, we should not expect that it is a good model for nature. For us the main purpose of this space is to develop an understanding of quantum field theory which has a broader range of applicability.

The noncommutative $\mathbb{R}^D$, $D = 2, 4, 6, \ldots$, is defined as the algebra $\mathbb{R}^D_\theta$ which as a vector space is given by the space $\mathcal{S}(\mathbb{R}^D)$ of (complex-valued) Schwartz class functions of rapid decay, equipped with the multiplication rule

$$(a \star b)(x) = \int \frac{d^D k}{(2\pi)^D} \int d^D y \; a(x+\tfrac{1}{2}\theta\cdot k)\, b(x+y)\, \mathrm{e}^{\mathrm{i}k\cdot y} \,, \tag{3}$$
$$(\theta\cdot k)^\mu = \theta^{\mu\nu} k_\nu \,, \quad k\cdot y = k_\mu y^\mu \,, \quad \theta^{\mu\nu} = -\theta^{\nu\mu} \,.$$

The entries $\theta^{\mu\nu}$ in (3) have the dimension of an area. The physical interpretation is $\|\theta\| \approx \ell_P^2$.

A field theory is defined by an action functional. We obtain action functionals on $\mathbb{R}^D_\theta$ by replacing in standard action functionals the ordinary product of functions by the $\star$-product. For example, the noncommutative ϕ^4-action is given by

$$S[\phi] := \int d^D x \Big(\frac{1}{2} \partial_\mu \phi \star \partial^\mu \phi + \frac{1}{2} m^2 \phi \star \phi + \frac{\lambda}{4!} \phi \star \phi \star \phi \star \phi \Big) \,. \tag{4}$$

The action (4) is then inserted into the partition function

$$Z[j] := \int \mathcal{D}\phi \; \mathrm{e}^{-\frac{1}{\hbar}(S[\phi] - \int d^4x\, \phi(x) j(x))} \,, \tag{5}$$

which gives rise to the correlation functions (= expectation values)

$$\langle\phi(x_1)\dots\phi(x_n)\rangle = Z[0]^{-1}\hbar^n \frac{\delta^n Z[j]}{\delta j(x_1)\dots\delta j(x_n)}\Big|_{j(x)=0} . \tag{6}$$

As usual we solve (5) perturbatively by Feynman graphs. Due to the fact that one star can be removed under the integral $\int d^D x\ (a\star b)(x) = \int d^D x\ a(x)b(x)$, the propagator in momentum space is unchanged.

The novelty are phase factors in the vertices, which reflect the cyclicity of the interaction integral,

$$\frac{\lambda}{4!}\,\mathrm{e}^{-\frac{i}{2}\sum_{i<j} p_i^\mu p_j^\nu \theta_{\mu\nu}} . \tag{7}$$

This leads to a convenient double line notation since the vertex contribution is invariant only under cyclic permutations of the legs (using momentum conservation). The resulting Feynman graphs are *ribbon graphs* which depend crucially on how the valences of the vertices are connected. For *planar graphs* the total phase factor of the integrand is independent of internal momenta, whereas *non-planar graphs* have a total phase factor which involves internal momenta. For example, the one-loop contribution to the two-point function splits as follows into a planar part and a non-planar part

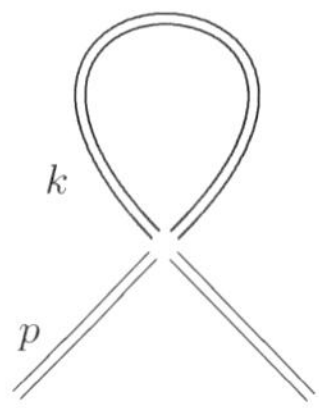

$$= \frac{\lambda}{6}\int dk \frac{1}{k^2+m^2}$$

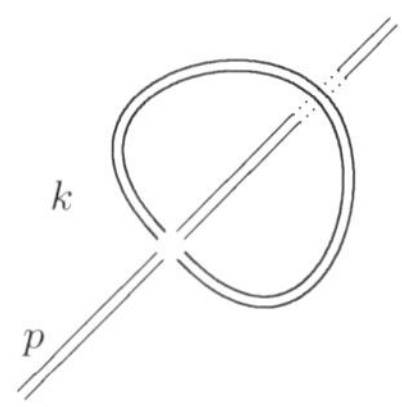

$$= \frac{\lambda}{12}\int dk \frac{\mathrm{e}^{ip^\mu k^\nu \theta_{\mu\nu}}}{k^2+m^2} \quad \overset{\sim}{p \to 0} \ \frac{1}{\tilde{p}^2}$$

where $\tilde{p}_\mu := \theta_{\mu\nu}p^\nu$.

Planar graphs are treated as usual. The resulting phase factor is precisely of the form of the original two-point function or vertex (7) so that the divergence can be removed via the normalization conditions. Here, the contribution can entirely be removed by a suitable normalization condition for the physical mass. The contribution from the non-planar graph is—at first sight—finite, which is a relict of the original motivation that noncommutativity would serve as regulator. The finiteness is important, because the momentum dependence does not appear in the original action (4), which means that a divergence of the form cannot be absorbed by multiplicative renormalisation.

However, the expansion of the modified Bessel function K_1 shows that the contribution behaves $\sim \tilde{p}^{-2}$ for small momenta. If we insert the nonplanar graph declared as finite as a subgraph into a bigger graph, one easily builds examples (with an arbitrary number of external legs) which leads to non-integrable integrals at small inner momenta. This is the so-called UV/IR-mixing problem.[4]

The heuristic argumentation can be made exact: Chepelev and Roiban have proven a power-counting theorem[5,6] which relates the power-counting degree of divergence to the topology of the ribbon graph. The rough summary of the power-counting theorem is that noncommutative field theories with quadratic divergences become meaningless beyond a certain loop order. The situation is better for field theories with logarithmic UV/IR-divergences, e.g. supersymmetric models. These can be formulated to any loop order. However, the logarithmic IR-divergences at exceptional external momenta are still present so that the correlation functions are unbounded: For every $\delta > 0$ one finds non-exceptional momenta such that $\big|\langle\phi(p_1)\ldots\phi(p_n)\rangle\big| > \frac{1}{\delta}$. In the next chapter we present an approach which solves these problems.

4. Renormalization

We have seen that quantum field theories on noncommutative $\mathbb{R}^D$ are not renormalisable by standard Feynman graph evaluations. One may speculate that the origin of this problem is the too naïve way one performs the continuum limit. A way to treat the limit more carefully is the use of flow equations. The idea goes back to Wilson. It was then used by Polchinski[7] to give a very efficient renormalisability proof of commutative ϕ^4-theory. Applying Polchinski's method to the noncommutative ϕ^4-model, we can hope to be able to prove renormalisability to all orders, too. There is, however, a serious problem of the momentum space proof. We have to guarantee that planar graphs only appear in the distinguished interaction coefficients for which we fix the boundary condition at the renormalisation scale Λ_R. Nonplanar graphs have phase factors which involve inner momenta. Polchinski's method consists in taking norms of the interaction coefficients, and these norms ignore possible phase factors. Thus, we would find that boundary conditions for non-planar graphs at Λ_R are required. Since there is an infinite number of different non-planar structures, the model is not renormalisable in this way. A more careful examination of the phase factors is also not possible because the cut-off integrals prevent the Gaußian integration required for the parametric integral representation.[5,6]

Fortunately, there is a matrix representation of the noncommutative $\mathbb{R}^D$, which we used in our first approach, where the $\star$-product becomes a simple product of infinite matrices. The price for this simplification is that the propagator becomes complicated, but the difficulties can be overcome.

4.1. *Matrix representation*

For simplicity we restrict ourselves to the noncommutative $\mathbb{R}^2$. There exists a matrix base $\{f_{mn}(x)\}_{m,n\in\mathbb{N}}$ of the noncommutative $\mathbb{R}^2$ which satisfies

$$(f_{mn}\star f_{kl})(x) = \delta_{nk} f_{ml}(x) , \qquad \int d^2x\, f_{mn}(x) = 2\pi\theta_1 , \tag{8}$$

where $\theta_1 := \theta_{12} = -\theta_{21}$. In terms of radial coordinates $x_1 = \rho\cos\varphi$, $x_2 = \rho\sin\varphi$ one has

$$f_{mn}(\rho,\varphi) = 2(-1)^m \mathrm{e}^{\mathrm{i}(n-m)\varphi}\sqrt{\tfrac{m!}{n!}}\left(\sqrt{\tfrac{2\rho^2}{\theta_1}}\right)^{n-m} L_m^{n-m}\left(\tfrac{2\rho^2}{\theta_1}\right)\mathrm{e}^{-\frac{\rho^2}{\theta_1}} , \tag{9}$$

where $L_n^\alpha(z)$ are the Laguerre polynomials. The matrix representation was also used to obtain exactly solvable noncommutative quantum field theories.[8,9]

Now we can write down the noncommutative ϕ^4-action in the matrix base by expanding the field as $\phi(x) = \sum_{m,n\in\mathbb{N}} \phi_{mn} f_{mn}(x)$. It turns out, however, that in order to prove renormalisability we have to consider a more general action than (4) at the initial scale Λ_0. This action is obtained by adding a harmonic oscillator potential to the standard noncommutative ϕ^4-action:

$$\begin{aligned} S[\phi] &:= \int d^2x \Big(\frac{1}{2}\partial_\mu\phi\star\partial^\mu\phi + 2\Omega^2(\tilde{x}^\mu\phi)\star(\tilde{x}_\mu\phi) + \frac{1}{2}\mu_0^2\phi\star\phi \\ &\qquad\qquad + \frac{\lambda}{4!}\phi\star\phi\star\phi\star\phi\Big)(x) \\ &= 2\pi\theta_1 \sum_{m,n,k,l}\Big(\frac{1}{2}G_{mn;kl}\phi_{mn}\phi_{kl} + \frac{\lambda}{4!}\phi_{mn}\phi_{nk}\phi_{kl}\phi_{lm}\Big) , \end{aligned} \tag{10}$$

where $\tilde{x}^\mu := \theta^{\mu\nu}x_\nu$ and

$$G_{mn;kl} := \int \frac{d^2x}{2\pi\theta_1}\Big(\partial_\mu f_{mn}\star\partial^\mu f_{kl} + 4\Omega^2(\tilde{x}^\mu f_{mn})\star(\tilde{x}_\mu f_{kl}) + \mu_0^2 f_{mn}\star f_{kl}\Big) . \tag{11}$$

We view Ω as a regulator and refer to the action (10) as describing a regularised ϕ^4-model. The action (10) could also be obtained by restricting a complex ϕ^4-model with magnetic field[8,9] to the real part. One finds

$$\begin{aligned} G_{mn;kl} &= \Big(\mu_0^2 + \frac{2}{\theta_1}(1+\Omega^2)(n+m+1)\Big)\delta_{nk}\delta_{ml} \\ &\quad - \frac{2}{\theta_1}(1-\Omega^2)\big(\sqrt{(n+1)(m+1)}\,\delta_{n+1,k}\delta_{m+1,l} - \sqrt{nm}\,\delta_{n-1,k}\delta_{m-1,l}\big)\,. \end{aligned} \tag{12}$$

The kinetic matrix $G_{mn;kl}$ has the important property that $G_{mn;kl} = 0$ unless $m + k = n + l$. The same relation is induced for the propagator $\Delta_{nm;lk}$ defined by $\sum_{k,l=0}^{\infty} G_{mn;kl}\Delta_{lk;sr} = \sum_{k,l=0}^{\infty} \Delta_{nm;lk}G_{kl;rs} = \delta_{mr}\delta_{ns}$. In order to evaluate the propagator we first diagonalise the kinetic matrix $G_{mn;kl}$:

$$G_{m,m+\alpha;l+\alpha,l} = \sum_{y\in\mathbb{N}} U_{my}^{(\alpha)}\big(\mu_0^2 + \tfrac{4\Omega}{\theta}(2y+\alpha+1)\big)U_{yl}^{(\alpha)}\,, \tag{13}$$

$$\begin{aligned} U_{ny}^{(\alpha)} &= \sqrt{\binom{\alpha+n}{n}\binom{\alpha+y}{y}\Big(\frac{1-\Omega}{1+\Omega}\Big)^{2n+2y+\alpha+1}\Big(\frac{4\Omega}{1-\Omega^2}\Big)^{\alpha+1}} \\ &\quad \times M_n\Big(y; 1+\alpha, \frac{(1-\Omega)^2}{(1+\Omega)^2}\Big)\,, \end{aligned} \tag{14}$$

where $M_n(y;\beta,c) = {}_2F_1\Big(\begin{smallmatrix} -n,-y \\ \beta \end{smallmatrix}\Big|1-c\Big)$ are the (orthogonal) Meixner polynomials. A lengthy calculation gives

$$\begin{aligned} \Delta_{mn;kl} &= \frac{\theta_1}{2(1+\Omega^2)}\delta_{m+k,n+l} \sum_{v=\frac{|m-l|}{2}}^{\frac{\min(m+l,k+n)}{2}} B\big(\tfrac{1}{2} + \tfrac{\mu_0^2\theta_1}{8\Omega} + \tfrac{1}{2}(m+k) - v, 1+2v\big) \\ &\quad \times \sqrt{\binom{n}{v+\frac{n-k}{2}}\binom{k}{v+\frac{k-n}{2}}\binom{m}{v+\frac{m-l}{2}}\binom{l}{v+\frac{l-m}{2}}}\Big(\frac{1-\Omega}{1+\Omega}\Big)^{2v} \\ &\quad \times {}_2F_1\left(\begin{matrix} 1+2v\,,\ \frac{1}{2} + \frac{\mu_0^2\theta_1}{8\Omega} - \frac{1}{2}(m+k)+v \\ \frac{3}{2} + \frac{\mu_0^2}{2\sqrt{1-\omega}\,\mu^2} + \frac{1}{2}(m+k)+v \end{matrix}\,\middle|\, \frac{(1-\Omega)^2}{(1+\Omega)^2}\right). \end{aligned} \tag{15}$$

Here, $B(a,b)$ is the Beta-function and $F(\begin{smallmatrix} a,b \\ c \end{smallmatrix};z)$ the hypergeometric function. We recall that in the momentum space version of the ϕ^4-model, the interactions contain oscillating phase factors which make a renormalisation by flow equations impossible. Here we use an adapted base which eliminates the phase factors from the interaction. We see from (15) that these oscillations do not reappear in the propagator. Note that all matrix elements $\Delta_{nm;lk}$ are non-zero for $m + k = n + l$.

4.2. *The Polchinski equation for matrix models*

We summarise here the derivation[10] of the Polchinski equation for the non-commutative ϕ^4-theory in the matrix base, which we derived with Raimar Wulkenhaar. According to Polchinski's derivation of the exact renormalisation group equation[7] we now consider a (at first sight) different problem than the matrix version of (5):

$$
\begin{aligned}
Z[J,\Lambda] &= \int \Big(\prod_{a,b} d\phi_{ab}\Big) \exp\big(-S[\phi,J,\Lambda]\big)\,,\\
S[\phi,J,\Lambda] &= (2\pi\theta_1)\Big(\sum_{m,n,k,l} \frac{1}{2}\phi_{mn} G^K_{mn;kl}(\Lambda)\,\phi_{kl} + \sum_{m,n,k,l} \phi_{mn} F_{mn;kl}[\Lambda] J_{kl}\\
&\qquad + \sum_{m,n,k,l} \frac{1}{2} J_{mn} E_{mn;kl}[\Lambda] J_{kl} + L[\phi,\Lambda] + C[\Lambda]\Big)\,,\\
G^K_{mn;kl}(\Lambda) &= \prod_{i\in\{m,n,k,l\}} K\Big[\tfrac{i}{\Lambda^2\theta_1}\Big]^{-1} G_{mn;kl}
\end{aligned}
\tag{16}
$$

with $L[0,\Lambda]=0$. The cut-off function $K(x)$ is a smooth decreasing function with $K(x)=1$ for $0\le x\le 1$ and $K(x)=0$ for $x\ge 2$. Accordingly, we define

$$
\Delta^K_{nm;lk}(\Lambda) = \prod_{i\in\{m,n,k,l\}} K\Big[\tfrac{i}{\Lambda^2\theta_1}\Big] \Delta_{nm;lk}\,. \tag{17}
$$

The function $C[\Lambda]$ is the vacuum energy and the matrices E and F, which are not necessary in the commutative case, must be introduced because the propagator Δ is non-local. It is in general not possible to separate the support of the sources J from the support of the Λ-variation of K. We would obtain the original problem for the choice

$$
\begin{aligned}
L[\phi,\infty] &= \sum_{m,n,k,l} \frac{\lambda}{4!}\phi_{mn}\phi_{nk}\phi_{kl}\phi_{lm}\,,\\
C[\infty] &= 0\,, \qquad E_{mn;kl}[\infty] = 0\,, \qquad F_{mn;kl}[\infty] = \delta_{ml}\delta_{nk}\,.
\end{aligned}
\tag{18}
$$

However, we shall expect divergences in the partition function which require a renormalisation, i.e. additional (divergent) counterterms in $L[\phi,\infty]$. In the Feynman graph solution of the partition function one carefully adapts these counterterms so that all divergences disappear. If such an adaptation is possible and all counterterms are local, the model is considered as perturbatively renormalisable.

Following Polchinski[7] we proceed differently to prove renormalisability. We first ask ourselves how we have to choose L, C, E, F in order to make

$Z[J,\Lambda]$ *independent* of Λ. After straightforward calculation one finds the answer

$$\Lambda\frac{\partial L[\phi,\Lambda]}{\partial\Lambda} = \sum_{m,n,k,l}\frac{1}{2}\Lambda\frac{\partial\Delta^K_{nm;lk}(\Lambda)}{\partial\Lambda}\bigg(\frac{\partial L[\phi,\Lambda]}{\partial\phi_{mn}}\frac{\partial L[\phi,\Lambda]}{\partial\phi_{kl}} - \frac{1}{2\pi\theta_1}\Big[\frac{\partial^2 L[\phi,\Lambda]}{\partial\phi_{mn}\,\partial\phi_{kl}}\Big]_\phi\bigg), \quad (19)$$

where $\big[f[\phi]\big]_\phi := f[\phi]-f[0]$. The corresponding differential equations for C,E,F are easy to integrate.[10] Now, instead of computing Green's functions from $Z[J,\infty]$ we can equally well start from $Z[J,\Lambda_R]$, where it leads to Feynman graphs with vertices given by the Taylor expansion coefficients $A^{(V)}_{m_1n_1;\dots;m_Nn_N}$ in

$$L[\phi,\Lambda] = \lambda\sum_{V=1}^{\infty}\big(2\pi\theta_1\lambda\big)^{V-1}\sum_{N=2}^{\infty}\frac{1}{N!}\sum_{m_i,n_i}A^{(V)}_{m_1n_1;\dots;m_Nn_N}[\Lambda]\phi_{m_1n_1}\cdots\phi_{m_Nn_N}\,. \quad (20)$$

These vertices are connected with each other by internal lines $\Delta^K_{nm;lk}(\Lambda)$ and to sources J_{kl} by external lines $\Delta^K_{nm;lk}(\Lambda_0)$. Since the summation variables are cut-off in the propagator (17), loop summations are finite, provided that the interaction coefficients $A^{(V)}_{m_1n_1;\dots;m_Nn_N}[\Lambda]$ are bounded. Thus, renormalisability amounts to prove that for certain initial conditions (parametrised by finitely many parameters!) the evolution of L according to (19) does not produce any divergences.

Inserting the expansion (20) into (19) and restricting to the part with N external legs we get the graphical expression. We see that for the simple fact that the fields ϕ_{mn} carry two indices, the effective action is expanded into ribbon graphs.

In the expansion of L there will occur very complicated ribbon graphs with crossings of lines which cannot be drawn any more in a plane. A general ribbon graph can, however, be drawn on a *Riemann surface* of some *genus* g. In fact, a ribbon graph *defines* the Riemann surfaces topologically through the *Euler characteristic* χ. We have to regard here the external lines of the ribbon graph as amputated (or closed), which means to directly connect the single lines m_i with n_i for each external leg m_in_i.

The genus is computed from the number $\tilde{L}$ of single-line loops, the number I of internal (double) lines and the number V of vertices of the graph according to Euler's formula $\chi = 2-2g = \tilde{L}-I+V$. The number

B of boundary components of a ribbon graph is the number of those loops which carry at least one external leg. There can be several possibilities to draw the graph and its Riemann surface, but $\tilde{L}, I, V, B$ and thus g remain unchanged. Indeed, the Polchinski equation (19) tells us which external legs of the vertices are connected. It is completely irrelevant how the ribbons are drawn between these legs. In particular, there is no distinction between overcrossings and undercrossings.

We expect that non-planar ribbon graphs with $g > 0$ and/or $B > 1$ behave differently under the renormalisation flow than planar graphs having $B = 1$ and $g = 0$. This suggests to introduce a further grading in g, B in the interactions coefficients $A^{(V,B,g)}_{m_1n_1;\ldots;m_Nn_N}$.

4.3. ϕ^4-theory on noncommutative $\mathbb{R}^2$

First one estimates the A-functions by integrating (19) perturbatively between an initial scale Λ_0 to be sent to ∞ later on and the renormalisation scale Λ_R:

Lemma 4.1. *The homogeneous parts* $A^{(V,B,g)}_{m_1n_1;\ldots;m_Nn_N}$ *of the coefficients of the effective action describing a regularised* ϕ^4*-theory on* $\mathbb{R}^2_\theta$ *in the matrix base are for* $2 \leq N \leq 2V+2$ *and* $\sum_{i=1}^N (m_i - n_i) = 0$ *bounded by*

$$\begin{aligned} &\left|A^{(V,B,g)}_{m_1n_1;\ldots;m_Nn_N}[\Lambda, \Lambda_0, \Omega, \rho_0]\right| \\ &\quad \leq \left(\Lambda^2\theta_1\right)^{2-V-B-2g}\left(\frac{1}{\Omega}\right)^{3V-\frac{N}{2}+B+2g-2} P^{2V-\frac{N}{2}}\left[\ln\frac{\Lambda_0}{\Lambda_R}\right]. \end{aligned} \tag{21}$$

We have $A^{(V,B,g)}_{m_1n_1;\ldots;m_Nn_N} \equiv 0$ *for* $N > 2V+2$ *or* $\sum_{i=1}^N (m_i - n_i) \neq 0$. *By* $P^q[x]$ *we denote a polynomial in* x *of degree* q.

The proof of (21) for general matrix models by induction goes over 20 pages! The formula specific for the ϕ^4-model on $\mathbb{R}^2_\theta$ follows from the asymptotic behaviour of the cut-off propagator (17), (15) and a certain index summation, see.[10,11]

We see from (21) that the only divergent function is

$$\begin{aligned} A^{(1,1,0)}_{m_1n_1;m_2n_2} &= A^{(1,1,0)}_{00;00}\delta_{m_1n_2}\delta_{m_2n_1} \\ &\quad + \left(A^{(1,1,0)}_{m_1n_1;m_2n_2}[\Lambda, \Lambda_0, \rho^0] - A^{(1,1,0)}_{00;00}\delta_{m_1n_2}\delta_{m_2n_1}\right), \end{aligned} \tag{22}$$

which is split into the distinguished divergent function

$$\rho[\Lambda, \Lambda_0, \Omega, \rho^0] := A^{(1,1,0)}_{00;00}[\Lambda, \Lambda_0, \Omega, \rho^0] \tag{23}$$

for which we impose the boundary condition $\rho_R := \rho[\Lambda_R, \Lambda_0, \Omega, \rho^0] = 0$ and a convergent part with boundary condition at Λ_0.

One remarks that the limit $\Omega \to 0$ in (21) is singular. In fact the estimation for $\Omega = 0$ with an optimal choice of the ρ-coefficients (different than (23)!) would grow with $\big(\Lambda\sqrt{\theta_1}\big)^{V-\frac{N}{2}-B-2g+2}$. Since the exponent of Λ can be arbitrarily large, there would be an infinite number of divergent interaction coefficients, which means that the ϕ^4-model is not renormalisable when keeping $\Omega = 0$.

In order to pass to the limit $\Lambda_0 \to \infty$ one has to control the total Λ_0-dependence of the functions $A^{(V,B,g)}_{m_1n_1;\dots;m_Nn_N}[\Lambda, \Lambda_0, \Omega[\Lambda_0], \rho^0[\Lambda_R, \Lambda_0, \rho_R]]$. This leads again to a differential equation in Λ, see.[11] It is then not difficult to see that the regularised ϕ^4-model with $\Omega > 0$ is renormalisable. It turns out that one can even prove more:[11] On can endow the parameter Ω for the oscillator frequency with an Λ_0-dependence so that in the limit $\Lambda_0 \to \infty$ one obtains a standard ϕ^4-model without the oscillator term:

Theorem 4.1. *The ϕ^4-model on $\mathbb{R}^2_\theta$ is (order by order in the coupling constant) renormalisable in the matrix base by adjusting the bare mass $\Lambda_0^2\rho[\Lambda_0]$ to give $A^{(1,1,0)}_{00;00}[\Lambda_R] = 0$ and by performing the limit $\Lambda_0 \to \infty$ along the path of regulated models characterised by $\Omega[\Lambda_0] = \big(1+\ln\frac{\Lambda_0}{\Lambda_R}\big)^{-1}$. The limit $A^{(V,B,g)}_{m_1n_1;\dots;m_Nn_N}[\Lambda_R, \infty] := \lim_{\Lambda_0\to\infty} A^{(V,B,g)}_{m_1n_1;\dots;m_Nn_N}[\Lambda_R, \Lambda_0, \Omega[\Lambda_0], \rho^0[\Lambda_0]]$ of the expansion coefficients of the effective action $L[\phi, \Lambda_R, \Lambda_0, \Omega[\Lambda_0], \rho^0[\Lambda_0]]$ exists and satisfies*

$$\begin{aligned}
&\Big|\lambda\big(2\pi\theta_1\lambda\big)^{V-1} A^{(V,B,g)}_{m_1n_1;\dots;m_Nn_N}[\Lambda_R, \infty] \\
&\qquad - \big(2\pi\theta_1\lambda\big)^{V-1} A^{(V,V^e,B,g,\iota)}_{m_1n_1;\dots;m_Nn_N}\Big[\Lambda_R, \Lambda_0, \tfrac{1}{(1+\ln\frac{\Lambda_0}{\Lambda_R})}, \rho^0[\Lambda_0]\Big]\Big| \\
&\leq \frac{\Lambda_R^4}{\Lambda_0^2}\Big(\frac{\lambda}{\Lambda_R^2}\Big)^V \Big(\frac{1+\ln\frac{\Lambda_0}{\Lambda_R}}{\Lambda_R^2\theta_1}\Big)^{B+2g-1} P^{5V-N-1}\Big[\ln\frac{\Lambda_0}{\Lambda_R}\Big]\,. \qquad (24)
\end{aligned}$$

In this way we have proven that the real ϕ^4-model on $\mathbb{R}^2_\theta$ is perturbatively renormalisable when formulated in the matrix base. This proof was not simply a generalisation of Polchinski's original proof to the noncommutative case. The naïve procedure would be to take the standard ϕ^4-action at the initial scale Λ_0, with Λ_0-dependent bare mass to be adjusted such that at Λ_R it is scaled down to the renormalised mass. Unfortunately, this does not work. In the limit $\Lambda_0 \to \infty$ one obtains an unbounded power-counting degree of divergence for the ribbon graphs. The solution is the observation that the cut-off action at Λ_0 is (due to the cut-off) not translation-invariant.

We are therefore free to break the translational symmetry of the action at Λ_0 even more by adding a harmonic oscillator potential for the fields ϕ. There exists a Λ_0-dependence of the oscillator frequency Ω with $\lim_{\Lambda_0\to\infty}\Omega = 0$ such that the effective action at Λ_R is convergent (and thus bounded) order by order in the coupling constant in the limit $\Lambda_0 \to \infty$. This means that the partition function of the original (translation-invariant) ϕ^4-model without cut-off and with suitable divergent bare mass can equally well be solved by Feynman graphs with propagators cut-off at Λ_R and vertices given by the bounded expansion coefficients of the effective action at Λ_R. Hence, this model is renormalisable, and in contrast to the naïve Feynman graph approach in momentum space[6] there is no problem with exceptional configurations.

4.4. ϕ^4-theory on noncommutative $\mathbb{R}^4$

The renormalisation of ϕ^4-theory on $\mathbb{R}^4_\theta$ in the matrix base is performed in an analogous way. We choose a coordinate system in which $\theta_1 = \theta_{12} = -\theta_{21}$ and $\theta_2 = \theta_{34} = -\theta_{43}$ are the only non-vanishing components of θ. Moreover, we assume $\theta_1 = \theta_2$ for simplicity. Then we expand the scalar field according to $\phi(x) = \sum_{m_1,n_1,m_2,n_2\in\mathbb{N}} \phi_{\substack{m_1\ n_1\\ m_2\ n_2}} f_{m_1n_1}(x_1,x_2) f_{m_2n_2}(x_3,x_4)$. The action (10) with integration over $\mathbb{R}^4$ leads then to a kinetic term generalising (12) and a propagator generalising (15). Using estimates on the asymptotic behaviour of that propagator one proves the four-dimensional generalisation of Lemma 4.1 on the power-counting degree of the N-point functions. For $\Omega > 0$ one finds that all non-planar graphs ($B > 1$ and/or $g > 0$) and all graphs with $N \geq 6$ external legs are convergent.

The remaining infinitely many planar two- and four-point functions have to be split into a divergent ρ-part and a convergent complement. Using some sort of locality for the propagator (15), which is a consequence of its derivation from Meixner polynomials, one proves that

A^{planar} are convergent functions, thus identifying

$$
\begin{aligned}
\rho_1 &:= A^{\text{planar}}_{\substack{0\,0\,;\,0\,0\\ 0\,0\,;\,0\,0}}\,,\\
\rho_2 &:= A^{\text{planar}}_{\substack{1\,0\,;\,0\,1\\ 0\,0\,;\,0\,0}} - A^{\text{planar}}_{\substack{0\,0\,;\,0\,0\\ 0\,0\,;\,0\,0}} = A^{\text{planar}}_{\substack{0\,0\,;\,0\,0\\ 1\,0\,;\,0\,1}} - A^{\text{planar}}_{\substack{0\,0\,;\,0\,0\\ 0\,0\,;\,0\,0}}\,,\\
\rho_3 &:= A^{\text{planar}}_{\substack{1\,1\,;\,0\,0\\ 0\,0\,;\,0\,0}} = A^{\text{planar}}_{\substack{0\,0\,;\,0\,0\\ 1\,1\,;\,0\,0}}\\
\rho_4 &:= A^{\text{planar}}_{\substack{0\,0\,;\,0\,0\,;\,0\,0\,;\,0\,0\\ 0\,0\,;\,0\,0\,;\,0\,0\,;\,0\,0}}
\end{aligned}
\tag{25}
$$

as the distinguished divergent ρ-functions for which we impose boundary conditions at Λ_R. Details are given in.[12]

The function ρ_3 has no commutative analogue. It corresponds to a normalisation condition for the frequency parameter Ω in (12). This means that in contrast to the two-dimensional case we cannot remove the oscillator potential with the limit $\Lambda_0 \to \infty$. In other words, the oscillator potential in (10) is a necessary companionship to the $\star$-product interaction. This observation is in agreement with the UV/IR-entanglement first observed in.[4] Whereas the UV/IR-problem prevents the renormalisation of ϕ^4-theory on $\mathbb{R}^4_\theta$ in momentum space,[6] we have found a self-consistent solution of the problem by providing the unique (due to properties of the Meixner polynomials) renormalisable extension of the action. We remark that the diagonalisation of the free action via the Meixner polynomials leads to discrete momenta as the only difference to the commutative case. The inverse of such a momentum quantum can be interpreted as the size of the (finite!) universe, as it is discussed in cosmology.

5. Taming the Landau ghost

After we found a way to cure the IR/UV problem it was a natural question to evaluate the β function, especially the question, whether the renormalization flow develops singularities, which run under the name of Landau ghosts and are related to the occurence of renormalons, which spoil Borel summability. The first order loop calculation[13] indicates that the model is not asymptotically free in the ultraviolet, as expected. On the other hand, to a certain surprise, the modification found due to switching on the oscillator revealed a zero of both beta functions at the Langmann-szabo self duality point at $\Omega = 1$, which indicates a new unexpected fixed point.

$$\lim_{\mathcal{N}\to\infty}\Big(\mathcal{N}\frac{\partial}{\partial\mathcal{N}} + N\gamma + \mu_0^2\beta_{\mu_0}\frac{\partial}{\partial\mu_0^2} + \beta_\lambda\frac{\partial}{\partial\lambda} + \beta_\Omega\frac{\partial}{\partial\Omega}\Big)\Gamma[\mu_0,\lambda,\Omega,\mathcal{N}] = 0 \tag{26}$$

$$\beta_\lambda = \frac{\lambda^2_{\text{phys}}}{48\pi^2}\frac{(1-\Omega^2_{\text{phys}})}{(1+\Omega^2_{\text{phys}})^3} + \mathcal{O}(\lambda^3_{\text{phys}})$$

$$\beta_\Omega = \mathcal{N}\frac{\partial}{\partial\mathcal{N}}\Big(\Omega[\mathcal{N}]\Big) = \frac{\lambda_{\text{phys}}\Omega_{\text{phys}}}{96\pi^2}\frac{(1-\Omega^2_{\text{phys}})}{(1+\Omega^2_{\text{phys}})^3} + \mathcal{O}(\lambda^2_{\text{phys}})$$

It is a remarkable fact, that the beta functions vanish at $\Omega = 1$.

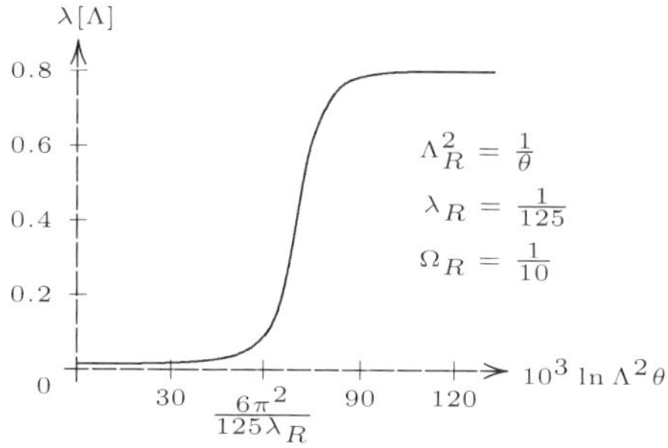

The flow equations can be solved easily. It turns out, that there is a fixed point of the RG flow at $\Omega = 1$.

As a next step the Paris group did calculate the beta function up to three loop[14] and the vanishing of the beta functions persisted.

In an interesting work it was possible to generalize this vanishing of the beta function to all loops by the Paris group in Reference.[15]

The reason behind this stabilization of the renormalization group flow lies in the fact, that there occurs a first order wave function renormalization and it follows that the flow becomes bounded, and the Landau ghost is (at least perturbatively) killed! This indicates the special properties of this model at the self-duality point, which might imply integrability. It might lead to a better understanding of such models. It is expected that they can be Borel summed.

As a summary we may say that $\Omega^2[\Lambda] \leq 1$, and $\lambda[\Lambda]$ is bounded, while in the commutative case $\lambda[\Lambda]$ diverges. It implies that

- the perturbation theory remains valid at all scales, and a
- non-perturbative construction of the model seems possible!

How does this work?

- four-point function renormalisation has the usual sign, but there
- $\exists$ an one-loop wave function renormalisation which compensates the four-point function renormalisation for $\Omega \to 1$

6. Induced Gauge Theory

There is a very simple and natural way to formulate gauge models on deformed spaces: We introduce covariant coordinates

$$\tilde{X}_\nu = \tilde{x}_\nu + A_\nu \tag{27}$$

whose motivation goes back to the transformation from the Euler to the Lagrangian picture in Hydrodynamics and require that they transform in the adjoint of unitary gauge transformation. The so introduced vector potential transforms than according to the following rule:

$$A_\mu \mapsto iu^* \star \partial_\mu u + u^* \star A_\mu \star u \tag{28}$$

It is now easy to introduce gauge invariant actions by replacing coordinates by covariant ones, for example the action of the noncommutative deformed Φ^4 model turns into the action treated in[16] and[17]

$$\begin{aligned} S = \int d^D x \bigg(& \frac{1}{2}\phi \star [\tilde{X}_\nu, [\tilde{X}^\nu, \phi]_\star]_\star + \frac{\Omega^2}{2}\phi \star \{\tilde{X}^\nu, \{\tilde{X}_\nu, \phi\}_\star\}_\star \\ & + \frac{\mu^2}{2}\phi \star \phi + \frac{\lambda}{4!}\phi \star \phi \star \phi \star \phi \bigg)(x) \end{aligned} \tag{29}$$

One considers now the vector potential as external and studies the heat kernel expansion of the scalar field theory coupled to this external potential. In both cited papers the very complicated and cumbersome one loop calculation has been done. Since the one-loop contributions are quadratically divergent one introduces an ultraviolet cutoff ϵ. In order to extract terms in the heat kernel expansion we used the Duhamel expansion leading to

$$\Gamma^\epsilon_{1l}\phi = -\frac{1}{2}\int_\epsilon^\infty \frac{dt}{t}\,\mathrm{Tr}\left(e^{-tH} - e^{-tH^0}\right) \tag{30}$$

The term proportional to $\frac{1}{\epsilon}$ gives a quadratic potential, the logarithmic divergent terms lead to an interesting proposal for an action of a gauge field on the noncommutative space

$$\begin{aligned} \Gamma^\epsilon_{1l} = \frac{1}{192\pi^2}\int d^4x \bigg\{ & \frac{24}{\tilde{\epsilon}\theta}(1-\rho^2)(\tilde{X}_\nu \star \tilde{X}^\nu - \tilde{x}^2) \\ & + \ln\epsilon \bigg(\frac{12}{\theta}(1-\rho^2)(\tilde{\mu}^2 - \rho^2)(\tilde{X}_\nu \star \tilde{X}^\nu - \tilde{x}^2) \\ & + 6(1-\rho^2)^2((\tilde{X}_\nu \star \tilde{X}^\nu)^{\star 2} - (\tilde{x}^2)^2) + \rho^4 F_{\mu\nu}F^{\mu\nu} \bigg) \bigg\}, \end{aligned} \tag{31}$$

where the field strength is defined through

$$F_{\mu\nu} = [\tilde{x}_\mu, A_\nu]_\star - [\tilde{x}_\nu, A_\mu]_\star + [A_\mu, A_\nu]_\star \tag{32}$$

The parameter ρ equals $frac1 - \omega^2 1 + \omega^2$ and becomes 1 for $\omega = 0$, it vanishes at the self-duality point.

Both terms of order $\mathcal{O}(1/\epsilon) + \mathcal{O}(\ln \epsilon)$ are gauge invariant. Of course it is not at all clear whether the so derived action is renormalizable? One drawback of this model is that under quatization the vacuum around $A = 0$ is not stable, a linear term in the action indicates a tadpole contribution, which should be taken into account. Some nontrivial vacuum solutions can be obtained, but quantization around these solution has not led to conclusions regarding renormalizability.

It is naturally to ask the question whether the BRST approach can be generalized to our noncommutative setting, and indeed the answer is yes. In common work with Daniel Blaschke and Manfred Schweda,[18] we were able to extend the quadratic part of the action including the oscillator term to a BRSR invariant action by introducing a vector ghost.

$$S = \int F^2/4 + \Omega^2/2 \int (\{\tilde{x}^\mu, A^\nu\} \star \{\tilde{x}_\mu, A_\nu\} + \{\tilde{x}_\mu \bar{c}, \tilde{x}^\mu c\}) \tag{33}$$

$$S_{gf} = \int B \star \partial_\mu A^\mu - B \star B/2 - \bar{c} \star \partial_\mu s A_\mu - \Omega^2 \tilde{c}_\mu \star s C_\mu \tag{34}$$

$$C_\mu = \{\{\tilde{x}_\mu \stackrel{\star}{,} A_\nu\} \stackrel{\star}{,} A_\nu\} + [\{\tilde{x}_\mu \stackrel{\star}{,} \bar{c}\} \stackrel{\star}{,} c] + [\bar{c} \stackrel{\star}{,} \{\tilde{x}_\mu \stackrel{\star}{,} c\}]$$

In conclusion: The total action including S_{gf} is BRST invariant. Invariant interaction terms can be added, of course. The BRST transformation is now given by

$$sA_\mu = D_\mu c, sc = igc \star c, s\tilde{c}_\mu, = \tilde{x}_\mu, sc = B, sB = 0 \tag{35}$$

and s squares to zero for all fields.

The main advantage of our action is that all propagators are now proportional to $(-\Delta + \Omega^2 \tilde{x}^2)^{-1}$, we can use the explicit form of this kernel using Mehler's formula. At present we are working out the one-loop calculation and expect no infrared-ultraviolet mixing to occur.

7. FERMIONS A spectral triple

The formulation of fermions is not so straightforward. A full treatment of a two dimensional model has been given by Vignes-Tourneret in.[19]

As for the four-dimensional case we tried to find a suitable square root of the Laplacian including the oscillator part. This turned out to be possible by observing that the oscillator in n dimensions can be understood as a spectral triple with spectral dimension n and KO-dimension equal to zero. In a recent work with Raimar Wulkenhaar[20] we extended this fact to the Moyal space. We take as a Dirac operator on Hilbert space $L^2(R^4) \otimes C^{16}$

$$D_8 = (i\Gamma^\mu \partial_\mu + \Omega \Gamma^{\mu+4} \tilde{x}_\mu) \tag{36}$$

where $\mu = 1, ...4$, and the matrices Γ_k generate the 8-dim Clifford algebra $\{\Gamma_k \Gamma_l\} = 2\delta_{kl}$. As it is seen, we take still a four dimensional space-time, but choose an eight dimensional Clifford algebra. The square of this Dirac operator becomes now

$$D_8^2 = (-\Delta + \Omega^2 ||\tilde{x}||^2) 1 - i\Omega \Theta^{-1}_{\mu\nu} [\Gamma^\mu, \Gamma^{\nu+4}]. \tag{37}$$

We may compute the action of this type of Dirac operator on sections of the spinor bundle and obtain

$$[D_8, f] * \psi = i[\Gamma^\mu + \Omega \Gamma^{\mu+4}](\partial \mu f) * \psi, \tag{38}$$

which means that only the four dimensional differential appears. Our configuration space dimension is still four, but the phase space dimensions becomes eight and equals the Clifford algebra dimension.
It turns out, that this dirac operator leads to a regular spectra triple in the sense of Connes.

8. Wedge-local QFT

After the formulation of quantum fields over deformed spaces, the question about a general formulation respectively modification of axioms of quantum fields has been addressed. The general attitude is of course, that Lorentz invariance and locality will be spoiled. In a common work with Gandalf Lechner[21] we found a formulation which allows to accept undeformed Lorentz symmetry and respects a special kind of locality, called wedge locality.

This work has been extended immediately by Buchholz and Summers (to be published).
We start with a formulation of free quantum fields over deformed Minkowski space time and choose as

noncommuting coordinates: $[\hat{x}^\mu, \hat{x}^\nu] = iQ^{\mu\nu}$, of the standard form

$$Q = \begin{pmatrix} 0 & \kappa_e & & \\ -\kappa_e & 0 & & \\ & & 0 & \kappa_m \\ & & -\kappa_m & 0 \end{pmatrix} \tag{39}$$

Field operators are defined as tensor products of Weyl operators times creation and/or annihilation operators:

$$\Phi_Q(x) = \int d\mu_p (e^{ipx} e^{ip\hat{x}} \otimes a_p^\dagger + e^{-ipx} e^{-ip\hat{x}} \otimes a_p) \tag{40}$$

where the deformed operators $a_{Q,p} = e^{-ip\hat{x}} \otimes a_p$ fulfill a twisted algebra, which I studied already in 1979 in connection with the quantization of integrable models and which is related to the so-called Zamolodchikov - Faddeev algebra.

$$a_{Q,p} a_{Q,p'} = e^{-ipQp'} a_{Q,p'} a_{Q,p}, \tag{41}$$

$$a_{Q,p} a_{Q,p'}^\dagger = e^{-ipQp'} a_{Q,p'}^\dagger a_{Q,p} + \omega_p \delta^{(3)}(\vec{p} - \vec{p'}) \tag{42}$$

Since we consider the family of field operators for the whole set of antisymmetric deformation matrices, there is no problem in obtaining the transformation properties under Lorentz transformations: The matrix Q just transforms as a tensor of second order and our model respects Lorentz symmetry. This is of course, similar to the method of Doplicher, Fredenhagen and Roberts.
The algebra of deformed creation/annihilation operators leads to twisted correlation functions,

$$\phi_Q(x_1)\ldots\phi_Q(x_N)|0> = \prod_{l<k} e^{-i\partial^\mu_{x_l} Q_{\mu\nu} \partial^\nu_{x_k}} \phi_0(x_1)\ldots\phi_0(x_N)|0> \tag{43}$$

similar to correlation functions obtained by Chaichian et al,[22] and Fiore and Wess[23]

Next we use a representation of fields on Hilbert space (H. Grosse 79)

$$A_Q(p) = e^{\frac{i}{2} pQ\hat{P}} a_p \tag{44}$$

where $\hat{P}$ is the momentum operator. A study of the properties of

$$\Phi_Q(x) = \int d\mu_p \Big(e^{ipx} A_{Q,p}^\dagger + e^{-ipx} A_{Q,p} \Big) \tag{45}$$

reveals that the correlation fucntions are tempered distribution, that the Reeh-Schlieder property holds, and of course, that they are not local, and not covariant. But as was mentioned already we treat all deformations and respect therefore Lorentz invariance. To generalize the construction we have to determine the algebra of $A_{Q,p}$ and $A^{\dagger}_{Q,p}$ for different Q using (44)

$$A_{Q,p}A_{Q',p'} = e^{-ip(Q+Q')/2p'}A_{Q',p'}A_{Q,p}, \tag{46}$$

$$A_{Q,p}A^{\dagger}_{Q',p'} = e^{-ip(Q+Q')/2p'}A^{\dagger}_{Q',p'}A_{Q,p} + \omega_p\delta^{(3)}(\vec{p}-\vec{p'})e^{ip(Q-Q')/2\hat{P}} \tag{47}$$

The usual Lorentz transformation properties result : It acts by the adjoint action on $A_{Q,p}$ and gives

$$U_{y,\Lambda}\Phi_Q(x)U^{\dagger}_{y,\Lambda} = \Phi_{\gamma_\Lambda(Q)}(\Lambda x + y), \;\; (y,\Lambda) \in \mathbb{P}$$

$$\gamma_\Lambda(Q) = \Lambda Q \Lambda^{\dagger}, \;\; \Lambda \in \mathcal{L}^{\uparrow}$$

As was mentioned, we treat all deformations and may therefore consider relative locality properties of fields $\Phi_Q(x)$. Especially we relate the antisymmetric matrices determining our family of fields to Wedges.

Wedges and Wedge local QF

We relate the antisymmetric matrices to Wedges: We start from the standard wedge: $W_1 = \left\{ x \in \mathbb{R}^D | x_1 > |x_0| \right\}$ and act on the standard wedge by proper Lorentz transformations $i_\Lambda(W) = \Lambda W$. The stabilizer group of the standard wedge is $SO(1,1)XSO(2)$, which corresponds to boosts and rotations.

$$\mathcal{W}_0 = \mathcal{L}^{\uparrow}_{+} W_1 \tag{48}$$

In addition we have to consider reflections: $j_\mu : x_\mu \mapsto -x_\mu$ $\mathcal{W}_0$ with $\mathcal{L}$ -action $i_\Lambda : \mathcal{W}_0 \mapsto \Lambda\mathcal{W}_0$ is a $\mathcal{L}_+$ - homogenous space. As a result we obtain an isomorphism between wedges and their transforms and the deformation matrices and their transforms $(\mathcal{W}_0, i_\Lambda) \cong (\mathcal{A}, \gamma_\Lambda)$ where

$\mathcal{A} = \{\gamma_\Lambda(Q_1) | \Lambda \in \mathcal{L}_+\}$ $Q(\Lambda W_1) := \gamma_\Lambda(Q_1)$

We define wedge local fields through: $\phi = \{\phi_W | W \subset \mathcal{W}_0\}$ and obtain a family of fields, which respect covariance and **localization in wedges**. With this isomorphism we define $\Phi_W(x) := \Phi_{Q(W)}(x)$. The transformation properties are of course given in the usual way

$$U_{y,\Lambda}\Phi_W(x)U^\dagger_{y,\Lambda} = \Phi_{\gamma_\Lambda(Q(W))}(\Lambda x + y) \tag{49}$$

and we obtain the

Theorem

Let $\kappa_e \geq 0$ the family $\Phi_W(x)$ is a wedge local quantum field on Fockspace:

$$[\phi_{W_1}(f), \phi_{-W_1}(g)](\psi) = 0,$$

for $supp(f) \subset W_1$, $supp(g) \subset -W_1$.
As for the proof we have to show that

$$[a_{Q_1}(f^-), a^\dagger_{-Q_1}(g^+)] + [a^\dagger_{Q_1}(f^+), a_{-Q_1}(g^-)] = 0 \tag{50}$$

Writing the integrals explicitly one can proof equ(refcomm) by doing an analytic continuation from R to $R + i\pi$ in the rapitity variable ϑ where $\vartheta = \sinh^{-1}\left(p_1/(m^2 + p_2^2 + p_3^2)^{1/2}\right)$

9. Conclusion

The subject of studies of renormalizability properties of deformed quantum field theories is quite young and only a few results have been obtained so far. Let us summarize a few of them:

- Removing cutoffs typically leads in ncQFT to IR/UV mixing, which signals that renormalizability will be spoiled in general. In some exceptional models only oriented graphs occur and one obtains no UV/IR problem.

- On eway to cure this disease consists in modifying the action for bosonic fields by adding one oscillator term (which breaks translation invariance). This leads to a renormalizable model.

- As a certain surprise the RG flow indicates a new fix point at the self duality point. It might lead to a nontrivial scalar Higgs model construction in the near future.

- This we can summarize by saying the renormalization group flows is save (bounded) and the will Landau ghost is tamed.

- Recently the Paris group found an additional way to cure the IR/UV problem[24] by adding a nonlocal counter term of the form $\frac{1}{p^2}$. In addition we were able in common work with Fabien Vignes-Tourneret to obtain a translational invariant version which consists in a kind of minimal version: We add one more nonlocal term and obtain a sensible renormalizable model.[25]

- We formulated also fermions and we expect that a regular spectral triple will result.
- As for the gauge fields we were able to formulate an action, and even an BRST invariant extension of this action has been obtained. Whether it will lead to renormalizability is still an open question.
- Of course one can hope that the final goal may be a noncommutative version of a renormalized noncommutative Standard model, which will not suffer from the triviality of the Higgs field.
- Finally a family of deformed fields has been considered on deformed Minkowski space-time. We were able to show that it fulfills at least WEDGE locality.
- Of course, there are many claims that these deformed fields include already gravity effects. Time will show how these toy models will teach us to learn more on approaches to quantize gravity.

10. Acknowledgments and Appendices

I would like to thank the organizers Prof. Erhard Seiler and Prof. Klaus Sibold for the kind invitation to this interesting meeting. I thank Prof. Raimar Wulkenhaar for an enjoyable collaboration over many years and my mathematical physics group (Gandalf Lechner, Karl-Georg Schlesinger, Harold Steinacker, Fabien Vignes-Tourneret, Michael Wohlgenannt and the PhD students) for many discussions.

References

1. S. Doplicher, K. Fredenhagen and J. E. Roberts, *Commun. Math. Phys.* **172**, 187 (1995).
2. S. Doplicher, K. Fredenhagen and J. E. Roberts, *Phys. Lett.* **B331**, 39 (1994).
3. A. Connes Noncommutative Geometry, Academic Press (1994).
4. S. Minwalla, M. Van Raamsdonk and N. Seiberg, *JHEP* **02**, p. 020 (2000).
5. I. Chepelev and R. Roiban, *JHEP* **05**, p. 037 (2000).
6. I. Chepelev and R. Roiban, *JHEP* **03**, p. 001 (2001).

7. J. Polchinski, *Nucl. Phys.* **B231**, 269 (1984).
8. E. Langmann, R. J. Szabo and K. Zarembo, *Phys. Lett.* **B569**, 95 (2003).
9. E. Langmann, R. J. Szabo and K. Zarembo, *JHEP* **01**, p. 017 (2004).
10. H. Grosse and R. Wulkenhaar, *Commun. Math. Phys.* **254**, 91 (2005).
11. H. Grosse and R. Wulkenhaar, *JHEP* **12**, p. 019 (2003).
12. H. Grosse and R. Wulkenhaar, *Commun. Math. Phys.* **256**, 305 (2005).
13. H. Grosse and R. Wulkenhaar, *Eur. Phys. J.* **C35**, 277 (2004).
14. M. Disertori and V. Rivasseau, *Eur. Phys. J.* **C50**, 661 (2007).
15. M. Disertori, R. Gurau, J. Magnen and V. Rivasseau, *Phys. Lett.* **B649**, 95 (2007).
16. H. Grosse and M. Wohlgenannt, *Eur. Phys. J.* **C52**, 435 (2007).
17. A. de Goursac, J.-C. Wallet and R. Wulkenhaar, *Eur. Phys. J.* **C51**, 977 (2007).
18. D. N. Blaschke, H. Grosse and M. Schweda, *Europhys. Lett.* **79**, p. 61002 (2007).
19. F. Vignes-Tourneret, *Annales Henri Poincare* **8**, 427 (2007).
20. H. Grosse and R. Wulkenhaar (2007).
21. H. Grosse and G. Lechner, *JHEP* **11**, p. 012 (2007).
22. M. Chaichian, P. P. Kulish, A. Tureanu, R. B. Zhang and X. Zhang (2007).
23. G. Fiore and J. Wess, *Phys. Rev.* **D75**, p. 105022 (2007).
24. R. Gurau, J. Magnen, V. Rivasseau and A. Tanasa (2008).
25. H. Grosse and F. Vignes-Tourneret (2008).

WARPED CONVOLUTIONS: A NOVEL TOOL IN THE CONSTRUCTION OF QUANTUM FIELD THEORIES

DETLEV BUCHHOLZ*

Institut für Theoretische Physik, Universität Göttingen
37077 Göttingen, Germany
**E-mail: buchholz@theorie.physik.uni-goettingen.de*

STEPHEN J. SUMMERS *

Department of Mathematics, University of Florida
Gainesville FL 32611, USA
**E-mail: sjs@math.ufl.edu*

Recently, Grosse and Lechner introduced a novel deformation procedure for non–interacting quantum field theories, giving rise to interesting examples of wedge–localized quantum fields with a non–trivial scattering matrix. In the present article we outline an extension of this procedure to the general framework of quantum field theory by introducing the concept of warped convolutions: given a theory, this construction provides wedge–localized operators which commute at spacelike distances, transform covariantly under the underlying representation of the Poincaré group and admit a scattering theory. The corresponding scattering matrix is nontrivial but breaks the Lorentz symmetry, in spite of the covariance and wedge–locality properties of the deformed operators.

Keywords: Quantum field theory; Constructive methods; Warped convolution

1. Introduction

Recent advances in algebraic quantum field theory have led to purely algebraic constructions of quantum field models on Minkowski space and other spacetimes, both classical and noncommutative,[2–5,8,11–15] many of which cannot be constructed by the standard methods of constructive quantum field theory. Some of these models are local and free, some are local and have nontrivial S-matrices, and yet others manifest only certain remnants of locality, though these remnants suffice to enable the computation of nontrivial two–particle S-matrix elements.

In a recent paper,[8] Grosse and Lechner have presented an infinite family of quantum fields which, taken as a whole, are wedge–local and Poincaré covariant and which have nontrivial scattering. They produce this family by deforming the free quantum field in a certain manner, motivated by the desire to understand the field as being defined on noncommutative Minkowski space. However, as they point out, one can forget the original motivation and view the resulting deformed fields as being defined on classical Minkowski space. It is, however, essential to the arguments of Ref. 8 that the *free* field is deformed.

In this paper we present a generalization of their deformation which can be applied to any Minkowski space quantum field theory in any number of dimensions. This deformation results in a one parameter family of distinct field algebras which are wedge–local and covariant under the representation of the Poincaré group associated with the initial, undeformed theory. It turns out that also the S–matrix changes under this deformation, and the deformed S–matrix breaks the Lorentz symmetry, in spite of the Lorentz covariance of the deformed theory. When taking the free quantum field as the initial model, our deformation coincides with that of Grosse and Lechner.

The deformation in question involves an apparently novel operator–valued integral, whose mathematical definition requires some care. Apart from the operators which are to be integrated, it involves a unitary representation of the additive group $\mathbb{R}^d$, $d \geq 2$, satisfying certain properties which arise naturally when considering relativistic quantum field theories on two (or higher) spacetime dimensional Minkowski space. We outline in Sec. 2 the intriguing properties of this integral; proofs will be given elsewhere. In Sec. 3 we apply these results to quantum field theories to obtain the results mentioned above. Finally, in Sec. 4 we indicate some paths of further investigation suggested by these results.

2. Warped convolutions

In order to draw attention to what may be regarded as the mathematical core of the deformation studied in this paper, we consider a quite general setting which covers both the case of Wightman Quantum Field Theory considered in Ref. 8 and the case of Algebraic Quantum Field Theory.[9]

We shall assume the existence of a strongly continuous unitary representation U of the additive group $\mathbb{R}^d$, $d \geq 2$, on some separable Hilbert space $\mathcal{H}$. The joint spectrum of the generators P of U is denoted by $\mathrm{sp}\, U$ and will be further specified in the following section. Let $\mathcal{D}$ be the dense subspace of vectors in $\mathcal{H}$ which transform smoothly under the action of U, *cf.* Ref. 6.

We consider the set $\mathfrak{F}$ of all operators F which have $\mathcal{D}$ in their domain of definition and are smooth under the adjoint action $\alpha_x(F) \doteq U(x)FU(x)^{-1}$ of U in the following sense: for each $F \in \mathfrak{F}$ there is some $n \in \mathbb{N}$ such that the operator valued function $x \mapsto (1+|P|^2)^{-n}\alpha_x(F)(1+|P|^2)^{-n}$ is arbitrarily often differentiable in norm, where $|P|^2$ denotes the sum of the squares of the generators of U. It is easily seen that $\mathfrak{F}$ is a unital *–algebra.

Within this framework one can establish a deformation procedure for the elements of $\mathfrak{F}$. The basic ingredient in this construction is the spectral resolution E of the unitary group U,

$$U(x) = e^{iPx} = \int e^{ipx}\, dE(p)\,, \quad x \in \mathbb{R}^d\,,$$

where the inner product on $\mathbb{R}^d$ is arbitrary here and will be fixed later. Given any skew–symmetric $d \times d$–matrix Q, *i.e.* $q\,Qp = -p\,Qq$ for $p, q \in \mathbb{R}^d$, one can give meaning to the operator valued integrals for any $F \in \mathfrak{F}$

$${}_QF \doteq \int \alpha_{Qp}(F)\, dE(p)\,, \quad F_Q \doteq \int dE(p)\, \alpha_{Qp}(F)\,. \tag{1}$$

These left and right integrals are defined on the domain $\mathcal{D}$ in the sense of distributions. Moreover, the resulting operators are smooth with regard to the adjoint action of U in the sense explained above; hence ${}_QF, F_Q \in \mathfrak{F}$. We omit the proof and only note that the above integrals may be viewed as warped (by the matrix Q) convolutions of F with the spectral measure dE.

The above integrals have a number of remarkable properties, which are crucial for their application to quantum field theory. We begin by noting the at first sight surprising fact that the left and right integrals coincide.

Lemma 2.1. *Let* $F \in \mathfrak{F}$. *Then* ${}_QF = F_Q$.

The proof of this lemma requires the proper treatment of expressions such as $dE(p)FdE(q)$ (which are not product measures) as well as the discussion of subtle domain problems. We therefore forego here a rigorous argument. Yet, in order to display the significance of the skew symmetry of the matrix Q for the result, we indicate the various steps in the proof. Making use several times of the relation $dE(p)f(P) = dE(p)f(p) = f(P)dE(p)$, which holds for any continuous function f, we get the following chain of equalities, which are justified below.

$$
\begin{aligned}
F_Q &= \int dE(p)\, \alpha_{Qp}(F) \\
&= \int dE(p)\, U(Qp) F U(Qp)^{-1} \int dE(q) \\
&= \iint dE(p)\, U(Qp) F\, U(Qp)^{-1}\, dE(q) \\
&= \iint dE(p)\, F e^{ipQq}\, dE(q) \\
&= \iint dE(p)\, e^{ipQq} F\, dE(q) \\
&= \iint dE(p)\, U(Qq) F U(Qq)^{-1}\, dE(q) \\
&= \int \alpha_{Qq}(F)\, dE(q) = {}_Q F\,.
\end{aligned}
$$

In the second equality we made use of $\int dE(q) = 1$, in the third one we relied on the fact that the preceding expression can be rewritten as a double integral, and in the fourth one we used the skew symmetry of Q, implying $dE(p)\, e^{-iPQp} = dE(p)$ and $e^{-iPQp}\, dE(q) = e^{-iqQp} dE(q) = e^{ipQq} dE(q)$. The fifth equality then follows, since e^{ipQq} is a c–number, and the sixth one is a consequence of $dE(p)\, e^{ipQq} = dE(p)\, e^{iPQq}$ and $dE(q) = e^{-iPQq}\, dE(q)$. In the last step we made use once again of $\int dE(p) = 1$.

It can be inferred from the defining relations (1) that $({}_Q F)^* \supset F^*{}_Q$. Thus, as an immediate consequence of the preceding lemma, one finds that the operation of taking adjoints commutes with the warped convolution in the following sense.

Lemma 2.2. *Let $F \in \mathfrak{F}$. Then ${F_Q}^* \supset F^*{}_Q$.*

It is also noteworthy that $(F_{Q_1})_{Q_2} = F_{Q_1+Q_2}$, for any $F \in \mathfrak{F}$ and skew symmetric matrices Q_1, Q_2. In the next lemma we exhibit commutation properties of certain specific elements of $\mathfrak{F}$, which are preserved by the deformation procedure. The shape of the spectrum $\mathrm{sp}\, U$ of the unitary group U, which coincides with the support of the spectral measure dE, enters in the formulation of this result.

Lemma 2.3. *Let $F, G \in \mathfrak{F}$ be such that*

$$\alpha_{Qp}(F)\, \alpha_{-Qq}(G) = \alpha_{-Qq}(G)\, \alpha_{Qp}(F)$$

for all $p, q \in sp\, U$. Then,

$$F_Q\, G_{-Q} = G_{-Q}\, F_Q\,.$$

Again, the rigorous proof of this result is plagued by technicalities and will not be given here. But the following formal steps, which are explained below, display the basic facts underlying the argument.

$$\begin{aligned}
F_Q\, G_{-Q} &= \int dE(p)\, \alpha_{Qp}(F) \int dE(q)\, \alpha_{-Qq}(G) \\
&= \int dE(p)\, \alpha_{Qp}(F) \int \alpha_{-Qq}(G)\, dE(q) \\
&= \iint dE(p)\, \alpha_{Qp}(F)\, \alpha_{-Qq}(G)\, dE(q) \\
&= \iint dE(p)\, \alpha_{-Qq}(G)\, \alpha_{Qp}(F)\, dE(q) \\
&= \iint dE(p)\, U(-Qq) G U(-Qq)^{-1}\, U(Qp) F U(Qp)^{-1}\, dE(q) \\
&= \iint dE(p)\, e^{-ipQq} G U(Qq + Qp) F e^{-iqQp}\, dE(q) \\
&= \iint dE(p)\, U(-Qp) G U(Qp + Qq) F U(-Qq)\, dE(q) \\
&= \iint dE(p)\, \alpha_{-Qp}(G)\, \alpha_{Qq}(F)\, dE(q) \\
&= \int dE(p)\, \alpha_{-Qp}(G) \int dE(q)\, \alpha_{Qq}(F) = G_{-Q}\, F_Q\,.
\end{aligned}$$

In the second equality use was made of Lemma 2.1, the third equality relies on the fact that the preceding product of operators can be presented as a double integral, and in the fourth equality the commutation properties of the operators F, G were exploited. The adjoint action of U is written out explicitly in the fifth equality, and in the sixth equality the group law for U as well as the relations $dE(p)\, e^{-iPQq} = dE(p)\, e^{-ipQq}$ and $e^{-iPQp}\, dE(q) = e^{-iqQp}\, dE(q)$ were used. The step to the seventh equality is accomplished by noting that the phase factors in the preceding expression cancel in view of the skew symmetry of Q, which also implies $dE(p) = dE(p)\, e^{-iPQp}$, $dE(q) = e^{-iPQq}\, dE(q)$. In the eighth equality the various unitaries are recombined into the form of adjoint actions, and in the subsequent equality the double integral is reexpressed as a product of simple integrals; Lemma 2.1 is then used once again.

We conclude this discussion of the warped convolution with a remark on its covariance properties. Let $\mathcal{L}$ be a matrix group acting isometrically (with regard to the chosen inner product) on $\mathbb{R}^d$ and let $\mathcal{P} = \mathcal{L} \ltimes \mathbb{R}^d$ be the semidirect product of the two groups. We assume that the unitary representation U of $\mathbb{R}^d$ can be extended to a representation of $\mathcal{P}$, denoted by the same symbol. Denoting the elements of $\mathcal{P}$ by $\lambda = (\Lambda, x)$, one then has $U(\lambda)U(y) = U(\Lambda y)U(\lambda)$ and consequently $U(\lambda)dE(p) = dE(\Lambda p)U(\lambda)$. It follows from standard arguments that $\mathfrak{F}$ is stable under the action of $\mathcal{P}$ given by $\alpha_\lambda(F) = U(\lambda)FU(\lambda)^{-1}$. Moreover,

$$U(\lambda)\Big(\int \alpha_{Qp}(F)\, dE(p)\Big)U(\lambda)^{-1} = \int \alpha_{\Lambda Qp}(U(\lambda)FU(\lambda)^{-1})\, dE(\Lambda p)$$
$$= \int \alpha_{\Lambda Q \Lambda^{-1} p}(U(\lambda)FU(\lambda)^{-1})\, dE(p)\,.$$

Note that the matrix $\Lambda Q \Lambda^{-1}$ is again skew symmetric with regard to the chosen inner product. We state the above result in the form of a lemma for later reference.

Lemma 2.4. *Let $F \in \mathfrak{F}$, let Q be any skew symmetric matrix and let $\lambda = (\Lambda, x)$ be any element of $\mathcal{P}$. Then*

$$\alpha_\lambda(F_Q) = \big(\alpha_\lambda(F)\big)_{\Lambda Q \Lambda^{-1}}\,.$$

With these results we have laid the foundation for the application of warped convolutions to quantum field theory.

3. Deformations of quantum field theories

We turn now to the discussion of local quantum field theories in Minkowski space and their deformations. Identifying d–dimensional Minkowski space with the manifold $\mathbb{R}^d$, the Lorentz inner product is given in proper coordinates by $xy = x_0 y_0 - \sum_{i=1}^{d-1} x_i y_i$. Any given quantum field theory on $\mathbb{R}^d$ may then be described as follows: there is a continuous unitary representation U of the Poincaré group $\mathcal{P} = \mathcal{L} \ltimes \mathbb{R}^d$ on a separable Hilbert space $\mathcal{H}$, where $\mathcal{L}$ is the identity component of the group of Lorentz transformations and $\mathbb{R}^d$ the group of spacetime translations. The joint spectrum of the generators P of the abelian subgroup $U \upharpoonright \mathbb{R}^d$ is contained in the closed forward lightcone $V_+ = \{p \in \mathbb{R}^d : p_0 \geq |\boldsymbol{p}|\}$ and there is a, up to a phase unique, unit vector $\Omega \in \mathcal{H}$, representing the vacuum, which is invariant under the action of U.

We assume that the underlying local field operators and observables generate a unital $*$–algebra $\mathfrak{A} \subset \mathfrak{F}$, where $\mathfrak{F}$ is the algebra of smooth operators with respect to the translations $U \upharpoonright \mathbb{R}^d$ introduced in the preceding section. In the Wightman setting of quantum field theory this assumption obtains if the underlying fields satisfy polynomial energy bounds.[6] In the framework of local quantum physics, where one deals with von Neumann algebras of bounded operators, one has to proceed to weakly dense subalgebras of elements smooth with respect to the action of the translation subgroup. So in both settings this assumption does not impose any significant restriction of generality and covers all models of interest.

The detailed structure of the theory is of no relevance here. What matters, however, is the assumption that one can identify all fields and observables which are localized in certain specific wedge–shaped regions, called wedges, for short. We fix a standard wedge

$$\mathcal{W}_0 \doteq \{x \in \mathbb{R}^d : x_1 \geq |x_0|\}$$

and note that in $d > 2$ dimensions all other wedges $\mathcal{W}$ can be obtained from $\mathcal{W}_0$ by suitable Poincaré transformations, $\mathcal{W} = \lambda\mathcal{W}_0$, $\lambda \in \mathcal{P}$. In $d = 2$ dimensions this statement only holds true if one also includes the spacetime reflections in $\mathcal{P}$.

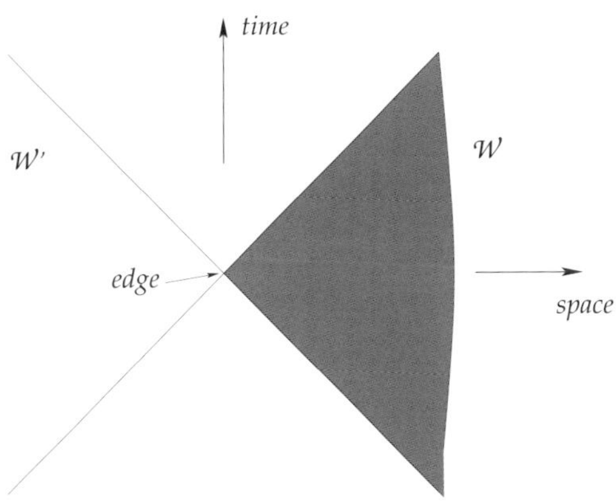

Fig. 1. A wedge $\mathcal{W}$, its causal complement $\mathcal{W}'$ and their common edge

Denoting by $\boldsymbol{\mathcal{W}} = \{\mathcal{W} \subset \mathbb{R}^d\}$ the set of all wedges in $\mathbb{R}^d$, we consider for any given $\mathcal{W} \in \boldsymbol{\mathcal{W}}$ the $*$–algebra $\mathfrak{A}(W) \subset \mathfrak{A}$ generated by all fields and observables localized in $\mathcal{W}$. We call the algebras $\mathfrak{A}(W)$ wedge–algebras. It is apparent from the definition that $\mathfrak{A}(W_1) \subset \mathfrak{A}(W_2)$ whenever $\mathcal{W}_1 \subset \mathcal{W}_2$, *i.e.* isotony holds. The covariance, locality and Reeh–Schlieder property of the underlying theory can then be expressed in terms of the wedge algebras as follows:

(a) *Covariance:* $\alpha_\lambda(\mathfrak{A}(\mathcal{W})) = U(\lambda)\mathfrak{A}(\mathcal{W})U(\lambda)^{-1} = \mathfrak{A}(\lambda\mathcal{W})$ for all $\mathcal{W} \in \boldsymbol{\mathcal{W}}$ and $\lambda \in \mathcal{P}$.
(b) *Locality:* $\mathfrak{A}(\mathcal{W}') \subset \mathfrak{A}(\mathcal{W})'$, $\mathcal{W} \in \boldsymbol{\mathcal{W}}$, where $\mathcal{W}'$ denotes the closure of the causal complement of $\mathcal{W}$ and $\mathfrak{A}(\mathcal{W})'$ the relative commutant of $\mathfrak{A}(\mathcal{W})$ in $\mathfrak{F}$.
(c) *Reeh–Schlieder property:* Ω is cyclic for any $\mathfrak{A}(\mathcal{W})$, $\mathcal{W} \in \boldsymbol{\mathcal{W}}$.

We mention as an aside that these assumptions also cover quantum field theories on non–commutative Minkowski space (Moyal space), as considered for example in Ref. 8. These spaces are described by non–commuting coordinates X_μ, X_ν satisfying the commutation relations

$$[X_\mu, X_\nu] = i\,\theta_{\mu\nu} 1\,,$$

where $\theta_{\mu\nu} = -\theta_{\nu\mu}$ are real constants. If the dimension of the spacetime satisfies $d > 2$, there exist lightlike coordinates $X_\pm$ with $[X_+, X_-] = 0$ which can thus be simultaneously diagonalized. Hence fields and observables on such spaces can be localized in wedges $\mathcal{W}$, yet they are dislocalized along the directions of the edges of these wedges. The wedge algebras are in general sufficient to reconstruct the algebras corresponding to arbitrary causally closed regions $\mathcal{R}$. These are given by

$$\mathfrak{A}(\mathcal{R}) \doteq \bigcap_{\mathcal{W} \supset \mathcal{R}} \mathfrak{A}(\mathcal{W})$$

and inherit from the wedge algebras both locality and covariance properties. Yet in theories on non–commutative Minkowski space, where fields and observables cannot be localized in bounded regions, the corresponding algebras are trivial and consequently do not manifest the Reeh–Schlieder property.

Given a theory as described above, we can now apply the deformation procedure established in the preceding section. To this end, we fix the standard wedge $\mathcal{W}_0$ and pick a corresponding $d \times d$–matrix Q_κ, which with respect to the chosen proper coordinates has the form

$$Q_\kappa \doteq \begin{pmatrix} 0 & \kappa & 0 & \cdots & 0 \\ \kappa & 0 & 0 & \cdots & 0 \\ 0 & 0 & 0 & \cdots & 0 \\ \vdots & \vdots & \vdots & \ddots & \vdots \\ 0 & 0 & 0 & \cdots & 0 \end{pmatrix}$$

for some fixed $\kappa > 0$. Note that this matrix is skew symmetric with respect to the Lorentz inner product. The following basic facts pointed out in Ref. 8 are crucial for the subsequent construction.

(i) Let $\lambda = (\Lambda, x) \in \mathcal{P}$ be such that $\lambda \mathcal{W}_0 \subset \mathcal{W}_0$. Then $\Lambda Q_\kappa \Lambda^{-1} = Q_\kappa$.
(ii) Let $\lambda' = (\Lambda', x') \in \mathcal{P}$ be such that $\lambda' \mathcal{W}_0 \subset \mathcal{W}_0{}'$. Then $\Lambda' Q_\kappa \Lambda'^{-1} = -Q_\kappa$.
(iii) $Q_\kappa V_+ = \mathcal{W}_0$.

It is an immediate consequence of (i) that for any two Poincaré transformations $\lambda_i = (\Lambda_i, x_i)$, $i = 1, 2$, such that $\lambda_1 \mathcal{W}_0 = \lambda_2 \mathcal{W}_0$, one has $\Lambda_1 Q_\kappa \Lambda_1^{-1} = \Lambda_2 Q_\kappa \Lambda_2^{-1}$. Indeed, $\lambda_2^{-1}\lambda_1 = \big(\Lambda_2^{-1}\Lambda_1, \Lambda_2^{-1}(x_1 - x_2)\big)$ maps $\mathcal{W}_0$ onto itself, hence $\Lambda_2^{-1}\Lambda_1 Q_\kappa \Lambda_1^{-1}\Lambda_2 = Q_\kappa$.

After these preparations we can now proceed from the given family of wedge algebras to a new "deformed" family with the help of the warped convolutions introduced in the preceding section. For $\mathcal{W} \in \boldsymbol{\mathcal{W}}$ the corresponding deformed algebras $\mathfrak{A}_\kappa(\mathcal{W})$ are defined as follows.

Definition 3.1. Let $\mathcal{W} \in \boldsymbol{\mathcal{W}}$ and let $\lambda = (\Lambda, x) \in \mathcal{P}$ be such that $\mathcal{W} = \lambda \mathcal{W}_0$. The associated algebra $\mathfrak{A}_\kappa(\mathcal{W})$ is the polynomial algebra generated by all warped operators $A_{\Lambda Q_\kappa \Lambda^{-1}}$ with $A \in \mathfrak{A}(\mathcal{W})$.

Note that according to the preceding remarks this definition is consistent, since it does not depend on the particular choice of the Poincaré transformation λ mapping $\mathcal{W}_0$ onto $\mathcal{W}$. Moreover, by Lemma 2.2, each $\mathfrak{A}_\kappa(\mathcal{W})$ is a $*$–algebra. We will show that the algebras $\mathfrak{A}_\kappa(\mathcal{W})$ have all desired properties of wedge algebras in a quantum field theory.

The isotony of the algebras $\mathfrak{A}_\kappa(\mathcal{W})$ is a consequence of the fact that if $\mathcal{W}_1 \subset \mathcal{W}_2$, these wedges can be mapped onto each other by a pure translation. Hence there are Poincaré transformations $\lambda_i = (\Lambda, x_i)$, $i = 1, 2$, with the same Λ mapping $\mathcal{W}_0$ onto $\mathcal{W}_1$ and $\mathcal{W}_2$, respectively. As $\mathfrak{A}_\kappa(\mathcal{W}_1)$, $\mathfrak{A}_\kappa(\mathcal{W}_2)$ are generated by the operators $A_{\Lambda Q_\kappa \Lambda^{-1}}$ with $A \in \mathfrak{A}(\mathcal{W}_1)$ and $A \in \mathfrak{A}(\mathcal{W}_2)$, respectively, the isotony of the original wedge algebras implies $\mathfrak{A}_\kappa(\mathcal{W}_1) \subset \mathfrak{A}_\kappa(\mathcal{W}_2)$ whenever $\mathcal{W}_1 \subset \mathcal{W}_2$.

For the proof of covariance we make use of Lemma 2.4. Let $\mathcal{W} = \lambda_{\mathcal{W}} \mathcal{W}_0$ with $\lambda_{\mathcal{W}} = (\Lambda_{\mathcal{W}}, x_{\mathcal{W}})$ and let $\lambda = (\Lambda, x)$. As the original theory is covariant, one has $\alpha_{\lambda_{\mathcal{W}}}(\mathfrak{A}(\mathcal{W}_0)) = \mathfrak{A}(\mathcal{W})$ and consequently the algebra $\mathfrak{A}_\kappa(\mathcal{W})$ is generated by the operators $\big(\alpha_{\lambda_{\mathcal{W}}}(A)\big)_{\Lambda_{\mathcal{W}} Q_\kappa \Lambda_{\mathcal{W}}{}^{-1}}$, $A \in \mathfrak{A}(\mathcal{W}_0)$. Now by Lemma 2.4

$$U(\lambda) \big(\alpha_{\lambda_{\mathcal{W}}}(A)\big)_{\Lambda_{\mathcal{W}} Q_\kappa \Lambda_{\mathcal{W}}{}^{-1}} U(\lambda)^{-1} = \big(\alpha_{\lambda\lambda_{\mathcal{W}}}(A)\big)_{\Lambda\Lambda_{\mathcal{W}} Q_\kappa \Lambda_{\mathcal{W}}{}^{-1}\Lambda^{-1}} ,$$

and the operators on the right hand side of this equality are, for $A \in \mathfrak{A}(\mathcal{W}_0)$, the generators of the algebra $\mathfrak{A}_\kappa(\lambda\mathcal{W})$. Thus $\alpha_\lambda(\mathfrak{A}_\kappa(\mathcal{W})) \subset \mathfrak{A}_\kappa(\lambda\mathcal{W})$. Replacing in this inclusion λ by λ^{-1} and $\mathcal{W}$ by $\lambda\mathcal{W}$ and making use of the fact that $\alpha_\lambda{}^{-1} = \alpha_{\lambda^{-1}}$, one obtains $\mathfrak{A}_\kappa(\lambda\mathcal{W}) \subset \alpha_\lambda(\mathfrak{A}_\kappa(\mathcal{W}))$. Hence $\alpha_\lambda(\mathfrak{A}_\kappa(\mathcal{W})) = \mathfrak{A}_\kappa(\lambda\mathcal{W})$, *i.e.* the deformed algebras satisfy the condition of covariance as well.

Turning to the proof of locality, we first restrict attention to the wedge $\mathcal{W}_0$. According to fact (iii) mentioned above, one has $Q_\kappa V_+ = \mathcal{W}_0$; hence $\mathcal{W}_0 + Q_\kappa p \subset \mathcal{W}_0$ and consequently $\mathcal{W}_0' \subset (\mathcal{W}_0 + Q_\kappa p)'$ for $p \in V_+$. Since V_+ is a cone, this implies $(\mathcal{W}_0' - Q_\kappa q) \subset (\mathcal{W}_0 + Q_\kappa p)'$ for all $p, q \in V_+$. It then follows from the covariance and locality properties of the original algebras that for any pair of operators $A \in \mathfrak{A}(\mathcal{W}_0)$ and $B \in \mathfrak{A}(\mathcal{W}_0')$ one has (denoting the pure translations $(1, x) \in \mathcal{P}$ by x)

$$[\alpha_{Q_\kappa p}(A), \alpha_{-Q_\kappa q}(B)] = 0\,, \quad p, q \in V_+\,.$$

According to Lemma 2.3, this implies $[A_{Q_\kappa}, B_{-Q_\kappa}] = 0$. Now if $\lambda = (\Lambda, x)$ is any Poincaré transformation such that $\lambda\mathcal{W}_0 = \mathcal{W}_0'$, it follows from fact (ii) mentioned above that $\Lambda Q_\kappa \Lambda^{-1} = -Q_\kappa$. Hence the operators B_{-Q_κ}, $B \in \mathfrak{A}(\mathcal{W}_0')$, generate the algebra $\mathfrak{A}_\kappa(\mathcal{W}_0')$, and similarly the operators A_{Q_κ}, $A \in \mathfrak{A}(\mathcal{W}_0)$, generate the algebra $\mathfrak{A}_\kappa(\mathcal{W}_0)$. So we obtain the inclusion $\mathfrak{A}_\kappa(\mathcal{W}_0') \subset \mathfrak{A}_\kappa(\mathcal{W}_0)'$. By the Poincaré covariance of the deformed algebras, established in the preceding step, it is then clear that $\mathfrak{A}_\kappa(\mathcal{W}') \subset \mathfrak{A}_\kappa(\mathcal{W})'$ for all $\mathcal{W} \in \boldsymbol{\mathcal{W}}$.

It remains to establish the Reeh–Schlieder property of the deformed algebras. According to Lemma 2.1, one has $A_Q = {}_Q A$ for any skew symmetric matrix Q. Hence $A_Q \Omega = {}_Q A \Omega = \int \alpha_{Qp}(A) dE(p) \Omega = A\Omega$, since Ω is invariant under spacetime translations. Thus $\mathfrak{A}_\kappa(\mathcal{W})\,\Omega \supset \mathfrak{A}(\mathcal{W})\,\Omega$ for any $\mathcal{W} \in \boldsymbol{\mathcal{W}}$, so the Reeh–Schlieder property of the deformed wedge algebras is inherited from the original algebras. We summarize these findings in a theorem.

Theorem 3.1. *Let $\mathfrak{A}(\mathcal{W}) \subset \mathfrak{F}$, $\mathcal{W} \in \boldsymbol{\mathcal{W}}$, be a family of wedge algebras having the Reeh–Schlieder property and satisfying the conditions of isotony, covariance, and locality. Then the family of deformed algebras $\mathfrak{A}_\kappa(\mathcal{W}) \subset \mathfrak{F}$, $\mathcal{W} \in \boldsymbol{\mathcal{W}}$, introduced in Definition 3.1 also has these properties.*

This theorem establishes that the deformation procedure outlined above can be applied to any quantum field theory. If one starts with the wedge algebras in a free field theory, one arrives at the deformed theories considered in Ref. 8, as can be seen by explicit computations. But one may equally well take as a starting point any rigorously constructed model, such as the self–interacting $\mathcal{P}(\varphi)$–theories in $d = 2$ dimensions or the φ^4–theory in $d = 3$ dimensions.[7] In all of these cases, the warped convolution produces a true deformation of the underlying theory, in the sense that the scattering matrix changes.

To exhibit this fact, let us assume that the underlying theory describes a single scalar massive particle. Then the spectrum of $U \restriction \mathbb{R}^d$ has the form

$$\operatorname{sp} U \restriction \mathbb{R}^d = \{0\} \cup \{p : p_0 = \sqrt{\boldsymbol{p}^2 + m^2}\} \cup \{p : p_0 \geq \sqrt{\boldsymbol{p}^2 + M^2}\},$$

with $M > m > 0$. In the present general setting of wedge–local operators one can then define two–particle scattering states as in Haag–Ruelle–Hepp scattering theory.[1] To see this, we fix the standard wedge $\mathcal{W}_0$ and pick operators $A \in \mathfrak{A}(\mathcal{W}_0)$ and $A' \in \mathfrak{A}(\mathcal{W}_0{}')$ which interpolate between the vacuum vector Ω and single particle states of mass m. We then proceed to the deformed operators $A_{Q_\kappa} \in \mathfrak{A}_\kappa(\mathcal{W}_0)$, $A'_{-Q_\kappa} \in \mathfrak{A}_\kappa(\mathcal{W}_0{}')$ and note that these operators have the same interpolation properties as the original ones, recalling that $A_{Q_\kappa}\Omega = A\Omega$, $A'_{-Q_\kappa}\Omega = A'\Omega$.

Next, we pick test functions $f, f' \in \mathcal{S}(\mathbb{R}^d)$ whose Fourier transforms $\widetilde{f}, \widetilde{f'}$ have compact supports in small neighborhoods of points on the isolated mass shell in $\operatorname{sp} U \restriction \mathbb{R}^d$ which do not intersect with the rest of the spectrum. With the help of these functions and the above operators we define

$$A_{Q_\kappa}(f_t) \doteq \int dx\, f_t(x)\, \alpha_x(A_{Q_\kappa}),$$

where the functions $f_t \in \mathcal{S}(\mathbb{R}^d)$, $t \in \mathbb{R}$, are given by

$$x \mapsto f_t(x) = (2\pi)^{-d/2} \int dp\, \widetilde{f}(p)\, e^{i(p_0 - \omega_{\boldsymbol{p}})t}\, e^{-ipx} \tag{2}$$

with $\omega_{\boldsymbol{p}} = (\boldsymbol{p}^2 + m^2)^{1/2}$. Similarly, one defines the operators $A'_{-Q_\kappa}(f'_t)$. Bearing in mind the support properties of $\widetilde{f}, \widetilde{f'}$ and the preceding remark about the action of the deformed operators on the vacuum vector, it follows that $A_{Q_\kappa}(f_t)\Omega = A(f_0)\Omega$ and $A'_{-Q_\kappa}(f'_t)\Omega = A'(f'_0)\Omega$ are single particle states which do not depend on t.

The operators $A_{Q_\kappa}(f_t)$, $A'_{-Q_\kappa}(f'_t)$ can be used to construct incoming, respectively outgoing, two–particle scattering states. Yet in the present case of wedge–localized operators this construction requires a proper adjustment of the support properties of the Fourier transforms of f, f'. Introducing the notation $\Gamma(g) \doteq \{(1, \boldsymbol{p}/\omega_{\boldsymbol{p}}) : p \in \operatorname{supp} \widetilde{g}\}$ for the velocity support of a test function g and writing $g_1 \succ g_2$ whenever the set $\Gamma(g_1) - \Gamma(g_2)$ is contained in the interior of the wedge $\mathcal{W}_0$, one relies on the following facts. According to a result of Hepp,[10] the essential supports of the functions f_t, f'_t are, for asymptotic t, contained in $t\,\Gamma(f)$, $t\,\Gamma(f')$, respectively. Moreover, the regions $\mathcal{W}_0 + t\Gamma(f)$ and $\mathcal{W}_0{}' + t\Gamma(f')$ are spacelike separated for $t < 0$ ($t > 0$) if $f' \succ f$ ($f \succ f'$), respectively. Because of the covariance and locality properties of the deformed wedge–algebras, one can then establish by standard arguments[1] the existence of the strong limits

$$\lim_{t\to-\infty} A_{Q_\kappa}(f_t)A'_{-Q_\kappa}(f'_t)\Omega \doteq |A(f)\Omega \otimes_\kappa A'(f')\Omega\rangle^{\text{in}} \quad \text{for } f' \succ f$$

$$\lim_{t\to\infty} A_{Q_\kappa}(f_t)A'_{-Q_\kappa}(f'_t)\Omega \doteq |A(f)\Omega \otimes_\kappa A'(f')\Omega\rangle^{\text{out}} \quad \text{for } f \succ f' \,.$$

The limit vectors have all properties of a symmetric tensor product of the single particle states $A(f)\Omega$, $A'(f')\Omega$. In particular, they do not depend on the specific choice of operators A, A' and test functions f, f' within the above limitations. Because of the Reeh–Schlieder property of the wedge algebras, it is also clear that these vectors form a basis in the respective asymptotic two–particle spaces.

In order to exhibit the dependence of the tensor products on the deformation parameter κ, we note that for $f' \succ f$

$$\begin{aligned}
|A(f)\Omega \otimes_\kappa A'(f')\Omega\rangle^{\text{in}} &= \lim_{t\to-\infty} A_{Q_\kappa}(f_t)A'_{-Q_\kappa}(f'_t)\Omega \\
&= \lim_{t\to-\infty} \int dE(p)\, \alpha_{Q_\kappa p}(A)(f_t)\, A'(f'_t)\Omega \\
&= \int dE(p)\, |U(Q_\kappa p)A(f)\Omega \otimes A'(f')\Omega\rangle^{\text{in}} \,,
\end{aligned}$$

where the third equality follows from the fact that the limit can be pulled under the integral and the symbol $\otimes$ denotes the tensor product in the original theory. Similarly, one obtains for $f \succ f'$

$$|A(f)\Omega \otimes_\kappa A'(f')\Omega\rangle^{\text{out}} = \int dE(p)\, |U(Q_\kappa p)A(f)\Omega \otimes A'(f')\Omega\rangle^{\text{out}} \,.$$

These relations between the scattering states in the original and in the deformed theory become more transparent if one proceeds to improper single particle states of sharp momentum $p = (\sqrt{\boldsymbol{p}^2 + m^2}, \boldsymbol{p})$, $q = (\sqrt{\boldsymbol{q}^2 + m^2}, \boldsymbol{q})$. There one has

$$|p \otimes_\kappa q\rangle^{\text{in}} = e^{i|pQ_\kappa q|} \, |p \otimes q\rangle^{\text{in}}$$
$$|p \otimes_\kappa q\rangle^{\text{out}} = e^{-i|pQ_\kappa q|} \, |p \otimes q\rangle^{\text{out}} .$$

The scattering states in the deformed theory depend on the matrix Q_κ through the choice of the wedge $\mathcal{W}_0$ and thus break the Lorentz symmetry in $d > 2$ dimensions. This can be understood if one interprets the wedge–local operators as members of a theory on non–commutative Minkowski space, where the Lorentz symmetry is broken.[8]

The kernels of the elastic scattering matrices in the deformed and undeformed theory are related by

$$^{out}\langle p \otimes_\kappa q | p' \otimes_\kappa q' \rangle^{in} = e^{i|pQ_\kappa q| + i|p' Q_\kappa q'|} \; ^{out}\langle p \otimes q | p' \otimes q' \rangle^{in} .$$

Thus they differ from each other, showing that the deformed and undeformed theories are not isomorphic. Yet since the difference is only a phase factor, the collision cross sections do not change under these deformations. Hence the effects of the deformation, such as the asymptotic breakdown of Lorentz invariance, could only be seen in certain specific arrangements such as time delay experiments.

4. Concluding remarks

In the present article we have presented a generalization of the deformation procedure of free quantum field theories, established by Grosse and Lechner,[8] to the general setting of relativistic quantum field theory. Even though the new theories which emerge in this way may not be of direct physical relevance, the results are of methodical interest. For they reveal yet again the significance of the wedge algebra in the algebraic approach to the construction of models.

From the algebraic point of view the problem of constructing a quantum field theory presents itself as follows. Given the stable particle content in the situation to be described, one first constructs a corresponding Fock space and representation U of the Poincaré group $\mathcal{P}$. A theory with this particle content is then obtained by fixing a wedge $\mathcal{W}_0$, say, and exhibiting a $*$–algebra $\mathfrak{G} \subset \mathfrak{F}$ which can be interpreted as the algebra generated by fields and observables localized in $\mathcal{W}_0$. It thus has to satisfy the conditions

(a) $\alpha_\lambda(\mathfrak{G}) \subset \mathfrak{G}$ whenever $\lambda \mathcal{W}_0 \subset \mathcal{W}_0$ for $\lambda \in \mathcal{P}$.
(b) $\alpha_{\lambda'}(\mathfrak{G}) \subset \mathfrak{G}'$ whenever $\lambda' \mathcal{W}_0 \subset \mathcal{W}_0'$ for $\lambda' \in \mathcal{P}$.

Any algebra $\mathfrak{G}$ satisfying these conditions is the germ of a quantum field theory in the following sense: setting $\mathfrak{A}(\mathcal{W}) \doteq \alpha_\lambda(\mathfrak{G})$, where $\lambda \in \mathcal{P}$ is such that $\mathcal{W} = \lambda\mathcal{W}_0$ for given $\mathcal{W} \in \boldsymbol{\mathcal{W}}$, it is an immediate consequence of the assumed properties of $\mathfrak{G}$ that the definition of the wedge algebras $\mathfrak{A}(\mathcal{W})$ is consistent and satisfies the conditions of isotony, covariance and locality. As explained above, the algebras corresponding to arbitrary causally closed regions can then consistently be defined by taking intersections of wedge algebras. Conversely, any asymptotically complete quantum field theory with the given particle content fixes an algebra $\mathfrak{G}$ with the above properties. Thus any quantum field theory can in principle be presented in this way. However, at present a dynamical principle by which the algebras $\mathfrak{G}$ can be selected is missing.

Nevertheless, this algebraic approach has already proven to be useful in the construction of interesting examples of quantum field theories. For instance, the existence of an infinity of models in $d = 2$ spacetime dimensions with non–trivial scattering matrix was established in this setting in Refs. 11–13, thereby solving a longstanding problem in the so–called form factor program of quantum field theory, *cf.* Ref. 15 and references quoted there. Wedge algebras associated with a nonlocal field in $d \geq 2$ spacetime dimensions were used in Ref. 5 to construct local observables manifesting non–trivial scattering. Wedge algebras were also used in Ref. 2 for the construction of quantum field theories describing massless particles with infinite spin, *cf.* also Ref. 14 for a construction of operators in these theories with somewhat better localization properties.

The idea of deforming given wedge algebras in order to arrive at new theories is a quite recent development in the algebraic approach and sheds new light on the constructive problems in quantum field theory. One may expect that the particular deformation procedure considered here is only an example of a richer family of similar constructions. Moreover, these methods can also be transferred to quantum field theories on curved spacetimes with a sufficiently big isometry group.

It is an intriguing question in this context to find manageable criteria which allow one to decide whether the intersections of wedge algebras are non–trivial. In Ref. 3 such a criterion based on the modular structure was put forward. Unfortunately, it is only meaningful in $d = 2$ spacetime dimensions. In the examples of deformed theories in $d > 2$ spacetime dimensions discussed here, it can be shown that the algebras corresponding to bounded spacetime regions are trivial. But, viewing the deformed theory as living on non–commutative Minkowski space,[8] one may expect that the algebras corresponding to the intersection of two opposite wedges are non–trivial. It would be of conceptual interest to establish this conjecture.

References

1. H.-J. Borchers, D. Buchholz and B. Schroer, Polarization–free generators and the S-matrix, *Commun. Math. Phys.*, ***219***, 125–140 (2001).
2. R. Brunetti, D. Guido and R. Longo, Modular localization and Wigner particles, *Rev. Math. Phys.*, ***14***, 759–785 (2002).
3. D. Buchholz and G. Lechner, Modular nuclearity and localization, *Ann. Henri Poincaré*, **5**, 1065–1080 (2004).
4. D. Buchholz and S. J. Summers, Stable quantum systems in Anti-de Sitter space: Causality, independence and spectral properties, *J. Math. Phys.*, ***45***, 4810–4831 (2004).
5. D. Buchholz and S. J. Summers, String– and brane–localized causal fields in a strongly nonlocal model, *J. Phys. A*, ***40***, 2147–2163 (2007).
6. K. Fredenhagen and J. Hertel, Local algebras of observables and pointlike localized fields, *Commun. Math. Phys.*, ***80***, 555–561 (1981).
7. J. Glimm and A. Jaffe, *Quantum Physics. A Functional Integral Point of View*, New York: Springer, 1987.
8. H. Grosse and G. Lechner, Wedge–local quantum fields and noncommutative Minkowski space, *JHEP*, **0711**, 012 (2007).
9. R. Haag, *Local Quantum Physics*, Berlin: Springer-Verlag, 1992.
10. K. Hepp: On the connection between Wightman and LSZ quantum field theory, pp. 135–246 in: *Brandeis University Summer Institute in Theoretical Physics 1965, "Axiomatic Field Theory"*, (M. Chretien and S. Deser eds.), Gordon and Breach 1966.
11. G. Lechner, Polarization-free quantum fields and interaction, *Lett. Math. Phys.*, ***64***, 137–154 (2003).
12. G. Lechner, On the existence of local observables in theories with a factorizing S-matrix, *J. Phys. A*, ***38***, 3045–3056 (2005).
13. G. Lechner, Construction of quantum field theories with factorizing S-matrices, *Commun. Math. Phys.*, **277**, 821–860 (2008).
14. J. Mund, B. Schroer and J. Yngvason, String–localized quantum fields and modular localization, *Commun. Math. Phys.*, ***268***, 621–672 (2006).
15. B. Schroer, Modular localization and the bootstrap–formfactor program, *Nucl. Phys. B*, ***499***, 547–568 (1997).

QUANTUM (OR AVERAGED) ENERGY INEQUALITIES IN QUANTUM FIELD THEORY

RAINER VERCH*

Institut für Theoretische Physik, Universität Leipzig
Vor dem Hospitaltore 1, D-04103 Leipzig, Germany
**E-mail: verch@itp.uni-leipzig.de*

A brief overview is given on the relation between energy conditions and spacetime geometry in solutions to the semiclassical Einstein equations of gravity. The quantum energy inequalities for Hadamard states of linear quantum fields on curved spacetimes are summarized. It is pointed out that quantum energy inequalities and the averaged null energy condition can be obtained for local thermal equilibrium states of general linear scalar quantum fields (including non-minimal curvature coupling) on globally hyperbolic spacetimes.

Keywords: Quantum field theory; Hadamard states; Energy inequalities

1. Introduction

The concept of a physical field, from its classical origins, emphasizes the the local character of a physical quantity in a system with infinitly many degrees of freedom. In the setting of local quantum physics, there are further characteristics and constraints on local quantities, like commutativity (only) at spacelike separation of field quantities, modelled as field operators, and positivity of the global energy of the system, together with existence of a ground state. These additional characteristics in cases render obstacles to a straightforward interpretation of local field operators in quantum field theory, and quite often they turn out to have unusual properties as compared to their classical counterparts, provided there are such. One of the early works attempting to clarify the meaning of local quantum field operators and to link them conceptually to stable elementary particles is the famous work by Lehmann, Symanzik and Zimmermann.[29] I read this work when I was a student, well before even thinking about what to do as a diploma thesis, out of a leisurely interest in quantum field theory, because I did not really understand at that time the way that field operators appeared in a course on elementary particle physics I took. Though I cannot really claim that I understood much of the LSZ work back those years, I was impressed by the conceptual clarity and rigour that this article was aiming at. In a sense, this work, together with the other by now classical texts on the all-model-embracing features of quantum field theory[20,26,35] made me wish to follow and better understand the subject, and it turned out that I continued to do, after all and at least, the former.

A particularly intriguing local quantity in physics, especially seen from the viewpoint of general relativity, is the stress-energy tensor of a field theory. To explain this, let us recall that in general relativity, the spacetime on which physical events can take place is mathematically described as a four-dimensional manifold M endowed with a Lorentzian (or semi-Riemannian) metric g_{ab}. (These terms are described, e.g., in the monograph[39] to which we refer the reader also for explanation of other concepts from Lorentzian and differential geometry appearing here and later in this text. We also adopt the abstract index notation for tensor fields on manifolds explained in that reference.) A spacetime is then formally given as a pair (M, g_{ab}). Associated with each spacetime are the covariant derivative ∇ of the metric g_{ab} and the corresponding curvature quantities, like the Ricci-tensor R_{ab} and its contraction $R = g^{ab}R_{ab}$, the scalar curvature. Out of these, one can build the Einstein-tensor $G_{ab} = R_{ab} - \frac{1}{2}g_{ab}R$. One of the fundamental principles of Einstein's theory of general relativity and gravity is that spacetime is not a fixed arena in which physical phenomena are staged, but rather, (M, g_{ab}) is an entity which is determined dynamically by Einstein's field equations[a]

$$G_{ab}(x) = 8\pi T_{ab}(x) \quad (x \in M)\,. \tag{1}$$

Here, T_{ab} is the stress-energy tensor of the matter and energy distributed in spacetime. For Einstein's equations, matter is modelled by a macroscopic phenomenological field theory, like electrodynamics or hydrodynamics. There are then also constraint equations (constitutive equations or propagation equations) describing the evolution of the field theoretical model. In this setting, Einstein's equations are formulated as an initial value problem: Given an initial spatial geometry described in terms of a 3-dimensional manifold Σ with a Riemannian metric h_{ab} on it (all space at "an instant of time"), put initial data for the matter field on Σ. Then, a solution to Einstein's equations consists of a spacetime (M, g_{ab}) and a matter field

[a]The field equations are here denoted in geometric units, cf.[39]

configuration with associated stress-energy tensor T_{ab} such that (1) holds together with the constraint equations for the matter model on all of M, and such that (Σ, h_{ab}) is a 3-dimensional slice at an instant of time of the 4-dimensional (M, g_{ab}) and with the restriction of the field configuration on M to Σ equalling the initial data. Thus, the geometry of spacetime is dynamically shaped by the evolution of matter distribution. On the other hand, since the curvature of spacetime geometry describes gravitational interaction (reflected in the constraint equations of the matter model) in the sense that point particles follow geodesics, the spacetime curvature determines also the evolution of the matter distribution in spacetime.

One may ask what the long-term fate of this interplay between spacetime geometry and matter distribution will turn out to be. Of course, the precise answer to that question is highly dependent on the specific form of the matter model and the initial data. There are also very subtle issues such as the stable dependence of solutions on the initial data, and it is not even very clear what (macroscopic) matter model (constraint equations) should be accepted as physical or how to assign a stress-energy tensor unambiguously to the matter model. However, some general statements of qualitative nature can be made. Macroscopic matter is stable – it does not decay spontaneously and there are no sinks for energy to leak away – and gravity is always attractive for macroscopic energy and matter. These basic experiences are, in fact, connected in general relativity. Let us suppose that to each matter models we consider, there is assigned a stress-energy tensor T_{ab}. Then, given any future-oriented timelike vector v^a with $v^a v_a = 1$, at any point $x \in M$, the quantity $T_{ab}(x)v^a v^b$ is the energy density of the matter model at x seen by an observer passing with proper velocity v^a through x. For many macroscopic matter models, it turns out that

$$T_{ab}(x)v^a v^b \geq 0 \quad \text{for all timelike vectors } v^a \text{ at } x \in M\,, \tag{2}$$

which means that the energy density of the matter model is positive for all observers in spacetime. Condition (2) is called the *weak energy condition*. If a stress-energy T_{ab} fulfills this condition (or related ones, cf. e.g.[22]) and appears on the right hand side of a solution to Einstein's equations, then the spacetime curvature on the left hand side of the solution is such that geodesics will tend to focus, i.e. the gravitational interaction described by the solution's spacetime geometry is always attractive. This is the central ingredient in order to derive conclusions about the long-term behaviour of solutions to Einstein's equations and it allows to conclude that under certain general conditions, it is inevitable that the spacetime geometry of a solution to Einstein's equations satisfying (2) will develop singularities in the future, or must have evolved from singularities in the past. (We refer to[22,28,39] for discussion.) Furthermore, the weak energy condition, or related conditions, basically rule out the development of a solution to Einstein's equations from a causally well-behaved spacetime geometry into a causally pathological spacetime scenario. The latter can include a spacetime geometry modelling the occurrence of closed timelike curves (which would correspond to the situation that an observer revisits her own past history, as though he were operating a time machine), or wormholes, which would allow a sort of superluminal travel by opening up "tunnels" to spatially remote regions.[30] Another scenario of superluminal travel are so-called "warp-drive" spacetime geometries.[1] The validity of arguments leading to cosmological singularities in Friedman-Robertson-Walker cosmological models is also questionable in case that weak energy conditions or their variants fail to hold.[32]

Having thus emphasized the prominent role of the stress-energy tensor as a local observable field quantity in the context of general relativity, let us turn to the features of this quantity in quantum field theory. In quantum field theory on Minkowski spacetime, we can consider a general quantum field theoretical model obeying Wightman's axioms,[35] and we may suppose that this quantum field theoretical model harbours also the stress-energy tensor as a local,covariant operator valued distribution which we denote by $\mathbf{T}_{ab}$. Under fairly general conditions, the operators $\mathbf{T}_{ab}(f)$, where f is a test function, converge to a quadratic form (in the sense of expectation values) when $f \to \delta_x$, i.e. if the test functions are being ideally peaked at the spacetime point x. Thus if ω is a "nice" state of the quantum field (bounded in energy), then the expectation value of stress-energy at x, $\langle \mathbf{T}_{ab}(x) \rangle_\omega = \lim_{f \to \delta_x} \langle \mathbf{T}_{ab}(f) \rangle_\omega$ is well-defined and smooth in x. Making the very reasonable assumption that the spatial integral of the energy density $\mathbf{T}_{00}(x)$ will yield the Hamiltonian at the level of expection values, it has been shown by Epstein, Glaser and Jaffe[9] that, at each spacetime point x, the quadratic form $\mathbf{T}_{00}(x)$ is not positive. For linear quantum field theories, it is even not too difficult to show that $\mathbf{T}_{00}(x)$ is unbounded (above and) below, meaning that there is, for each spacetime point x, a sequence of nice states ω_n so that $\langle \mathbf{T}_{00}(x) \rangle_{\omega_n} \to -\infty$ as $n \to \infty$. A similar conclusion can be drawn for general Wightman fields with a certain short distance scaling behaviour; we refer to the lucid discussion of these matters in a very nice report by Chris Fewster.[10] This result has, though indirectly, experimental manifestations demonstrating the occurrence of negative energy densities due to quantum effects. Among the important examples of this are the Casimir effect, and squeezed states of light.

Seen again from the perspective of general relativity, negative energy densities may have important consequences regarding the long-term behaviour of spacetime geometry when quantum effects of matter are taken into account. A way of modelling the quantum nature of matter interacting with geometry is provided by the *semiclassical Einstein equations*

$$G_{ab}(x) = 8\pi \langle \mathbf{T}_{ab}(x) \rangle_\omega \,. \tag{3}$$

Let us briefly discuss this equation. We assume that for every spacetime (M, g_{ab}) we are given a quantum field theory on that spacetime (satisfying, among other things, the principle of local general covariance, see[25] for a recently proposed general setup, and also the next section), and each of those theories possesses a local stress-energy quantum field $\mathbf{T}_{ab}$ of which expectation values $\langle \mathbf{T}_{ab}(x) \rangle_\omega$ can be formed at each spacetime point $x \in M$ for a set of "nice" states ω. A solution to (3) is then a spacetime (M, g_{ab}) together with a state ω of the quantum field theory on (M, g_{ab}) such that (3) holds. The basic assumption underlying the semiclassical Einstein equations is that spacetime geometry can still be described "classically", i.e. without inclusion of potential quantum effects into the description of spacetime geometry, even in regimes where quantum effects are no longer negligible in the description of matter. Since the interaction strengths of elementary particle processes are many orders of magnitude larger than that of gravity, one believes that there is actually a regime involving very high curvatures of spacetime geometry where this approximation is valid. As such, the said basic assumption, and the semiclassical Einstein equations, are to be seen as a semiclassical approximation to a full theory of quantum gravity – since such theory is not available to date, such a semiclassical approximation may serve as an important guideline for finding essential ingredients of quantum gravity. The Hawking effect,[21] predicting thermal radiation by black holes due to quantum effects on the matter side, is an example of a situation where quantum effects of matter are important, but where spacetime geometry is still described classically. Taking the validity of the semiclassical approximation for granted, one may now wonder about the long-term time evolution of solutions to (3). A priori one would expect that this might be quite different from the typical long-term time evolution of solutions to the classical Einstein equations in view of the fact that the expected energy density $\langle \mathbf{T}_{ab}(x) \rangle_\omega v^a v^b$ appearing on the right hand side of the the semiclassical Einstein equations can be made as negative as desired. This implies that geodesics won't focus but diverge in solutions to (3), provided that negativity of the expected energy density is strong enough and sustained over a sufficiently large domain in spacetime. All sorts of causally pathological behaviour of solutions to the semiclassical Einstein equations is then possible in principle.

2. Quantum Fields on Curved Spacetime

It turns out that there are constraints on the amount and duration of negative expected negative energy densities in quantum field theory. Essentially, these constraints can be traced back on the principle of stability of matter. In order to be more specific, let us now introduce a very simple model for a generally covariant quantum field theory illustrating matters here. Take a four-dimensional spacetime (M, g_{ab}) (other dimensions would do just as well) and assume that this spacetime possesses a time-orientation and is globally hyperbolic. Global hyperbolicity means that M can be sliced by Cauchy-surfaces, where a Cauchy-surface is a 3-dimensional submanifold which is intersected exactly once by each inextendible causal curve in M. Thus, a Cauchy-surface is a submanifold of events at the same instant of time, and serves as a submanifold on which initial data for hyperbolic differential equations can be freely posed. With the spacetime (M, g_{ab}) we associate the Klein-Gordon operator

$$K = \nabla^a \nabla_a + \xi R + m^2$$

acting on smooth scalar (real-valued) funtions on M. Here, ∇ is the covariant derivative of the metric g_{ab} and R is its scalar curvature, while ξ and m are real parameters, usually assumed to be non-negative. Then

$$K\varphi = 0\,, \quad (\varphi \in C^\infty(M, \mathbb{R}))$$

is the scalar wave equation on (M, g_{ab}). Owing to the assumption of global hyperbolicity, there are unique advanced and retarded fundamental solutions, or Green's functions, E^{adv} and E^{ret} for K, defined on $C_0^\infty(M, \mathbb{R})$. Their difference $E = E^{\text{adv}} - E^{\text{ret}}$ is called causal Green's function. We introduce now a $*$-algebra $\mathcal{F}(M, g_{ab})$ generated by a unit element $\mathbf{1}$ and symbols $\phi(f)$, $f \in C_0^\infty(M, g_{ab})$, with the relations

$$f \mapsto \phi(f) \text{ is linear}\,, \quad \phi(f)^* = \phi(f) \ (f \text{ real})\,,$$
$$\phi(Kf) = 0\,, \, [\phi(f), \phi(h)] = i\langle f, Eh\rangle \cdot \mathbf{1}\,.$$

For the last condition, $[A, B] = AB - BA$ denotes the commutator, and $\langle f, Eh\rangle = \int\int f(x)E(x,y)h(y)d\mu(x)d\mu(y)$ in formal notation for distributions, where $d\mu$ is the spacetime volume form induced by the metric g_{ab}.

Note that the last condition is the covariant form of the canonical commutation relations. Thus, $\mathcal{F}(M, g_{ab})$ is an abstract version of the "quantized linear scalar field" model on the spacetime (M, g_{ab}). Now we have an algebra of abstract field operators $\phi(f)$, and what is needed in addition are states on this algebra and Hilbert-space representations. A state is a linear functional $\langle\,.\,\rangle_\omega : \mathcal{F}(M, g_{ab}) \to \mathbb{C}$ which is positive, meaning $\langle A^*A\rangle_\omega \geq 0$ for all $A \in \mathcal{F}(M, g_{ab})$, normalized by $\langle \mathbf{1}\rangle_\omega = 1$, and continuous in the sense that the maps $f_1, \ldots, f_n \mapsto \langle\phi(f_1)\cdots\phi(f_n)\rangle_\omega$ are distributions. Associated with each state $\langle\,.\,\rangle_\omega$ is a Hilbert-space representation of $\mathcal{F}(M, g_{ab})$, denoted $(\mathcal{H}_\omega, \pi_\omega, \Omega_\omega)$, and referred to as GNS-representation, where $\mathcal{H}_\omega$ is the representation Hilbert-space, π_ω is a representation of $\mathcal{F}(M, g_{ab})$ by closable operators on the common dense domain domain $\pi_\omega(\mathcal{F}(M, g_{ab}))\Omega_a \subset \mathcal{H}_\omega$ so that $\pi_\omega(A^*)$ agrees with the restriction of $\pi_\omega(A)^*$ to the domain, where Ω_ω is a unit vector in $\mathcal{H}_\omega$ with the property that $(\pi_\omega(A)\Omega_\omega, \pi_\omega(B)\Omega_\omega) = \langle A^*B\rangle_\omega$ (with the scalar product in $\mathcal{H}_\omega$ on the left hand side) for all $A, B \in \mathcal{F}(M, g_{ab})$.

It is well known that not every "state" according to this mathematical definition corresponds to a physically realistic configuration of the system "quantized linear scalar field". There are pathological "states" modelling infinite particle densities, infinite temperatures and the like. Such "states" would not be regarded as physical states of the system, and it is necessary to select the physical states by means of suitable criteria. A central criterion is energetical stability (expressing stability of matter), and one initial step in an attempt to formulate that more formally is to demand that there should be a reasonably good definition of the expected stress-energy tensor $\langle \mathbf{T}_{ab}(x)\rangle_\omega$ for each physical state $\langle\,.\,\rangle_\omega$. This condition is less trivial than it might appear at first sight since there is no element in $\mathcal{F}(M, g_{ab})$ which would correspond in any sense to a quantized version of the stress-energy tensor of the linear scalar field. In fact, the definition of $\langle \mathbf{T}_{ab}(x)\rangle_\omega$ inevitably involves a process of infinite renormalization. Note also that one cannot rely on a selection criterion based on a distinguished behaviour of certain states with respect to space-time symmetries since generic globally hyperbolic spacetimes need not possess any symmetries. Nevertheless, the selection criterion for physical states on a generic spacetime has to include, for reasons of consistency, distinguished states with respect to spacetime symmetries, such as vacua or thermal equilibrium states, if the underlying spacetime admits corresponding spacetime symmetries. To say that a definition of $\langle \mathbf{T}_{ab}(x)\rangle_\omega$ is "reasonably good" would need also a definition, but some conditions to this effect can in fact be stated quite naturally. A first condition is local general covariance. Suppose that there are two globally hyperbolic spacetimes, (M, g_{ab}) and (M', g'_{ab}), and suppose that these posesses globally hyperbolic sub-spacetimes (in the most stringent interpretation of "globally hyperbolic sub-spacetime") (N, g_{ab}) and (N', g'_{ab}), respectively, where the restrictions of g_{ab} to N and g'_{ab} to N' haven't been indicated by extra symbols but are implicitly understood. Suppose further there is a bijective isometry $\psi : N \to N'$. [b] In this case, one can show that there is a bijective $*$-algebraic morphism $\alpha_\psi : \mathcal{F}(N, g_{ab}) \to \mathcal{F}(N', g'_{ab})$, canonically induced by $\alpha_\phi(\phi(f)) = \phi'(f \circ \psi^{-1})$. Moreover, if this situation is iterated, with a globally hyperbolic sub-spacetime (N'', g''_{ab}) of another globally hyperbolic spacetime (M'', g''_{ab}), and a bijective isometry $\chi : (N', g'_{ab}) \to (N'', g''_{ab})$, then it holds that $\alpha_\chi \circ \alpha_\psi = \alpha_{\psi \circ \chi}$. This is the formal expression of saying that the assignment of algebras $\mathcal{F}(M, g_{ab})$ to globally hyperbolic spacetimes (M, g_{ab}) fulfills the principle of local general covariance.[4,37] To say that the definition of $\langle \mathbf{T}_{ab}(x) \rangle_\omega$ is locally covariant then means that, in the situation where $\psi : N \to N'$ is a bijective isometry, it holds that

$$\langle \mathbf{T}_{ab}(x) \rangle_{\alpha^*_\psi \omega'} = \psi_* \langle \mathbf{T}'_{ab}(\psi(x)) \rangle_{\omega'} \quad (x \in N)$$

for all physical states ω' on $\mathcal{F}(M', g'_{ab})$. Here, $\langle \mathbf{T}_{ab}(x) \rangle_\omega$ is the renormalized expected stress-energy tensor defined for physical states ω on $\mathcal{F}(M, g_{ab})$, and $\langle \mathbf{T}'_{ab}(x) \rangle_{\omega'}$ is the like quantity defined for physical states ω' on $\mathcal{F}(M', g'_{ab})$; α^*_ψ is the dual of α_ψ, defined by $\langle A \rangle_{\alpha^*_\psi \omega'} = \langle \alpha_\psi(A) \rangle_{\omega'}$ for $A \in \mathcal{F}(M, g_{ab})$. A second condition for a "reasonably good" definition of $\langle \mathbf{T}_{ab}(x) \rangle_\omega$ is to demand that it behaves correctly with respect to spacetime symmetries, at least in the following sense. If the spacetime (M, g_{ab}) is static, there is a timelike Killing vector field ξ^a which is orthogonal to a Cauchy-surface Σ in the spacetime. ξ^a generates a 1-parametric group of time-translations τ_t $(t \in \mathbb{R})$ on M. They lift to automorphisms α_t on $\mathcal{F}(M, g_{ab})$ by $\alpha_t(\phi(f)) = \phi(f \circ \tau_{-t})$. If n^a denotes the normalization of ξ^a to unit length, then $\langle \mathbf{T}_{ab} n^a n^b(x) \rangle_\omega$ is the expected energy density on Σ, and its integral should generate the time-evolution,

$$\int_\Sigma \langle [\mathbf{T}_{ab} n^a n^b(x), A] \rangle_\omega d\mu_\Sigma(x) = -i \left. \frac{d}{dt} \right|_{t=0} \langle \alpha_t(A) \rangle_\omega ,$$

where $d\mu_\Sigma$ denotes the volume form induced on Σ. Finally, it is generally viewed as desirable that the expected stress-energy tensor be divergence-free, $\nabla^a \langle \mathbf{T}_{ab}(x) \rangle_\omega = 0$.

---

[b]In this paper, an isometry is always assumed to preserve orientation and time-orientation.

What we have described here corresponds to (part of) the requirements on the $\langle \mathbf{T}_{ab}(x)\rangle_\omega$ for physical states ω originally formulated by Wald,[38,40] which motivated the definition of "Hadamard states". For a Hadamard state $\langle\,.\,\rangle_\omega$, its two-point function $\langle \phi(f)\phi(h)\rangle_\omega$ has a singular part which is determined by the spacetime geometry and the field equation for ϕ and is thus state-independent. In some more detail, for a Hadamard state one has

$$\langle \phi(x)\phi(y)\rangle_\omega = \frac{U(x,y)}{\sigma(x,y)} + V(x,y)\ln(\sigma(x,y)) + W_\omega(x,y) \qquad (4)$$

where $\sigma(x,y)$ is the squared geodesic distance between spacetime points x and y, a $U(x,y)$ and $V(x,y)$ are functions determined by the spacetime geometry and the field equation for ϕ, and $W_\omega(x,y)$ is a smooth function containing the dependence of the two-point function on the state ω. Of course, (4) is an oversimplification in several respects. First, the two-point function is a distribution, and one needs to supply a prescription in which sense the singularities are to be treated (an "$i\epsilon$" prescription) since $\sigma(x,y)$ is zero if the points x and y coincide or can be connected by a lightlike geodesic. Furthermore, $\sigma(x,y)$ is only locally well-defined, and the same applies to $U(x,y)$ and $V(x,y)$. The function $V(x,y)$ is even only defined as an asymptotic power series in $\sigma(x,y)$ where the coefficients are determined by the Hadamard recursion relations. We shall not go into further detail on these matters here; suffice it to say that they have been settled in a rigorous manner in the literature.[27] The basic idea of defining $\langle \mathbf{T}_{ab}(x)\rangle_\omega$ for Hadamard states, following,[38,40] is then as follows. For the classical real scalar field $\varphi(x)$, the energy momentum tensor $T_{ab}(x)$ at spacetime point x is given as

$$\begin{aligned} T_{ab}(x) &= (\nabla_a\varphi(x))(\nabla_b\varphi(x)) + \frac{1}{2}g_{ab}(x)(m^2\varphi^2(x) - (\nabla^c\varphi)(\nabla_c\varphi)(x)) \\ &\quad + \xi(g_{ab}(x)\nabla^c\nabla_c - \nabla_a\nabla_b - G_{ab}(x))\varphi^2(x)\,. \end{aligned}$$

Now one defines first $W^s_\omega(x,y) = \frac{1}{2}(W_\omega(x,y) + W_\omega(y,x))$, and defines

$$\begin{aligned} \langle :\phi\nabla_a\phi:(x)\rangle_\omega &= \nabla_{a'}W^s_\omega\lfloor_x \\ \langle :\phi\nabla_a\nabla_b\phi:(x)\rangle_\omega &= \nabla_{a'}\nabla_{b'}W^s_\omega\lfloor_x \\ \langle :(\nabla_a\phi)(\nabla_b\phi):(x)\rangle_\omega &= \nabla_a\nabla_{b'}W^s_\omega\lfloor_x\,. \end{aligned}$$

Here, $\nabla_{a'}W^s_\omega\lfloor_x$ means applying ∇ on $W^s_\omega(x,x')$ with respect to x', and taking the coincidence limit $x' \to x$ afterwards. In the coincidence limit, primed and unprimed tensor indices are identified. With this definition, one sets

$$\begin{aligned}\langle \mathbf{T}_{ab}(x)\rangle_\omega &= -\langle : \phi\nabla_a\nabla_b\phi : (x) + \frac{1}{4}\nabla_a\nabla_b : \phi^2 : (x)\rangle_\omega \\ &+ \left(\frac{1}{4} - \xi\right) \langle \nabla_a\nabla_b : \phi^2 : (x) - g_{ab}(x)\nabla^c\nabla_c : \phi^2 : (x)\rangle_\omega \\ &+ \frac{1}{2}g_{ab}(x)\langle : \phi\nabla^c\nabla_c\phi : (x) + m^2 : \phi^2 : (x)\rangle_\omega \\ &- \xi G_{ab}(x)\langle : \phi^2 : (x)\rangle_\omega - Q(x)g_{ab}(x)\end{aligned}$$

A term $-Q(x)g_{ab}(x)$ has been added to render $\langle \mathbf{T}_{ab}(x)\rangle_\omega$ divergence-free. (See[31] for an alternative approach.) The function Q is also constructed locally from the spacetime geometry. Moreover, a Leibniz-rule for derivatives of the Wick squares $: \phi^2 : (x)$ has been exploited, see[34] for details. This definition can then be shown to be "reasonable" according to the criteria mentioned above (in particular, the definition is generally covariant). Note that the infinite renormalization comes about by discarding the singular, geometry-determined part of the two-point function.

There remains some renormalization ambiguity which is not ruled out by the criteria on a "reasonable" expected stress-energy tensor. Adding a tensor field C_{ab} to the above $\langle \mathbf{T}_{ab}(x)\rangle_\omega$ results in a re-definition of the renormalized expected stress-energy tensor which is still in agreement with the above criteria as long as C_{ab} is locally constructed from the metric and divergence-free. A typical form of C_{ab} is

$$C_{ab} = A_1 g_{ab} + A_2 G_{ab} + A_3 \frac{\delta}{\delta g^{ab}} S_3(g) + A_4 \frac{\delta}{\delta g^{ab}} S_4(g) \tag{5}$$

where $S_3(g) = \int_M R^2 \mathrm{dvol}_g$, $S_3(g) = \int_M R_{ab}R^{ab}\mathrm{dvol}_g$, and $\delta/\delta g^{ab}$ means functional differentiation with respect to the metric. Invoking scaling arguments, this ambiguity of the renormalized stress-energy tensor is already the general form, so that the remaining renormalization ambiguity resides in the four free constants A_1, A_2, A_3, A_4. We refer the reader to[38,40] for further discussion on this point.

We should mention that the property of a state's two-point function to be of Hadamard form can be equivalently expressed by the requirement that the two-point function have a wave-front set of a particular form.[33,36] This important result by Radzikowski has had considerable influence on the development of quantum field theory in curved spacetime. It was observed that conditions on the wavefront sets of the n-point functions of states on $\mathcal{F}(M, g_{ab})$ can also be imposed in a fairly natural manner – these conditions have been called microlocal spectrum conditions – and they have been instrumental in defining local and covariant normal ordered and time ordered products of the linear scalar field on curved spacetime, and hence in the local and covariant perturbative construction of scalar, self-interacting quantum field theories in generic curved spacetimes.[3,23,24] We recommend that the reader takes a closer look at this important development which we can only briefly mention in passing.

3. Quantum Energy Inequalities

Now that a characterization of physical states of the linear scalar field on curved spacetime has been given at least for the purpose of defining "reasonable" expectation values of the stress-energy tensor, we can try and see what information can be gained about the failure of the expected energy density $\langle \mathbf{T}_{ab}(x)\rangle_\omega v^a v^b$ to be positive for timelike vectors v^a at spacetime points x. As already remarked, this quantity, at each x, is unbounded above and below as a functional on the Hadamard states. But matters change when passing from the pointwise quantity to averaged quantities. L. Ford argued in[17] that spacetime averages of the expected stress-energy, in particular of the expected energy density, are unlikely to become arbitrarily negative for long duration of averaging, since this could lead to macroscopic violations of the second law of thermodynamics. This issue was thence investigated by Ford and others for free quantized fields on Minkowski spacetime[18] which confirmed his proposal and leads to what has then been called "quantum inequalities", but the more recent term "quantum energy inequalities" appears preferable since it is less ambiguous. To be more specific, suppose that $\mathcal{L}$ is a set of states on $\mathcal{F}(M, g_{ab})$ for a globally hyperbolic spacetime (M, g_{ab}). We assume that this set is contained in the set of Hadamard states so that the (renormalized) expected stress-energy tensor is well-defined. Then we say that this set of states fulfills a *timelike averaged quantum energy inequality* if, for every smooth, timelike curve γ in M with proper time parameter t and tangent $\dot\gamma^a$, and for each weighting function $h \in C_0^\infty(\mathbb{R})$, there is a constant $q(\gamma, h) > -\infty$ such that

$$\int_{-\infty}^{\infty} \langle \mathbf{T}_{ab}(\gamma(t))\rangle_\omega \dot\gamma^a(t)\dot\gamma^b(t)|h(t)|^2\, dt \geq q(\gamma, h) \qquad (6)$$

holds for all states $\langle\,.\,\rangle_\omega \in \mathcal{L}$. This means, in other words, that for given timelike curve and weighting function, the weighted integral of the expected energy density along the curve is bounded below and the lower bound is independent of the states in the set $\mathcal{L}$. One can change the definition by replacing "timelike curve" by "lightlike curve", and correspondingly one obtains the condition on $\mathcal{L}$ to fulfill a *lightlike averaged quantum energy inequality.* If the conditions are only required for timelike (or lightlike) *geodesics* γ, then one says that $\mathcal{L}$ fulfills a *timelike (lightlike) geodesic quantum energy inequality.* Henceforth we will adopt the common abbreviation QEI for quantum energy inequality.

The validity of QEIs for linear quantum fields on curved spacetime has been intensively investigated over the last decade, with the majority of rigorous results attached to the name of Chris Fewster. A first important result by Fewster was the proof that the set of all Hadamard states of the quantized minimally coupled Klein-Gordon field (corresponding to curvature coupling $\xi = 0$ in the definition of the wave operator K), on arbitrary globally hyperbolic spacetimes, fulfills a timelike QEI. This result on the validity of a timelike QEI for all Hadamard states was then extended to the quantized Dirac field and the quantized free electromagnetic field on globally hyperbolic spacetimes.[11] Somewhat surprisingly, a lightlike QEI was found not to hold for all Hadamard states in spacetime dimension 4.[15] It may also be a bit unexpected that timelike QEIs do not hold for the non-minimally coupled quantized linear scalar field (where $\xi \neq 0$ in the definition of K), even in Minkowski spacetime. Nevertheless, it has been shown that the violation of energy positivity for the non-minimally coupled case is also restricted, by so-called *relative quantum energy inequalities.* Here, one says that a set $\mathcal{L}$ of states $\langle\,.\,\rangle_\omega$ fulfills a relative QEI if, given a timelike curve γ and C_0^∞ weighting function h, there is a quadratic form $Q(\gamma, h)$ on $\mathcal{L}$ so that

$$\int_{-\infty}^{\infty} \langle \mathbf{T}_{ab}(\gamma(t))\rangle_\omega \dot{\gamma}^a(t)\dot{\gamma}^b(t)|h(t)|^2\,dt \geq \langle Q(\gamma,h)\rangle_\omega$$

holds for all states $\langle\,.\,\rangle_\omega \in \mathcal{L}$. The quadratic form $Q(\gamma, h)$ can, in this case, be unbounded (if $Q(\gamma, h)$ is bounded, one recovers QEIs in the case already discussed), but of course a relative QEI should be nontrivial, and this requires that for any choice of positive constants C and C', the estimate

$$C|\langle Q(\gamma,h)\rangle_\omega| + C' \geq \int_{-\infty}^{\infty} \langle \mathbf{T}_{ab}(\gamma(t))\rangle_\omega \dot{\gamma}^a(t)\dot{\gamma}^b(t)|h(t)|^2\,dt$$

is *violated* for some states $\langle\,.\,\rangle_\omega$ in $\mathcal{L}$. For the case of the non-minimally coupled linear scalar field on Minkowski spacetime, Fewster and Osterbrink have established a timelike averaged relative QEI for Hadamard states where the quadratic form $Q(\gamma, h)$ is essentially the number operator.[13]

Another topic are *absolute* QEIs. The terminology suggests that these were in some contrast to relative QEIs, but actually "absolute" refers to something else here, namely the absence of a dependence of the lower bounds $q(\gamma, h)$ on a choice of reference state. One says that a set of states $\mathcal{L}_{(M,g_{ab})}$ on $\mathcal{F}(M, g_{ab})$ fulfills an absolute quantum energy inequality if the constant $q(\gamma, h) = q_{(M,g_{ab})(\gamma,h)}$ in (6) depends only on the parameters defining the quantum field model (i.e. m, ξ, and e.g. the constants A_1, A_2, A_3, A_4 determining C_{ab} in (5)), and depends on the spacetime geometry only in a local and covariant manner. That is to say, if $\psi : (M, g_{ab}) \to (M', g'_{ab})$ is an isometry, then $\alpha^*_\psi(\mathcal{L}_{(M',g'_{ab})}) = \mathcal{L}_{(M,g_{ab})}$ and $q_{(M',g'_{ab})}(\psi \circ \gamma, h) = q_{(M,g_{ab})}(\gamma, h)$. Absolute QEIs have been proved for Hadamard states of the minimally coupled scalar field;[16] see also[14] for further discussion.

There are also other variants of averages of the expected energy density along causal curves which can be viewed as certain limiting cases of QEIs. A particular example is the *averaged null energy condition* (ANEC). One says that a set of states $\mathcal{L}$ on $\mathcal{F}(M, g_{ab})$ fulfills the ANEC if for each complete lightlike geodesic γ in (M, g_{ab}) one has that

$$\liminf_{\lambda \to 0} \int_{-\infty}^{\infty} \langle \mathbf{T}_{ab}(\gamma(t)) \rangle_\omega \dot{\gamma}^a(t) \dot{\gamma}^b(t) \eta(\lambda t)\, dt \geq 0 \tag{7}$$

holds for all states $\langle\,.\,\rangle_\omega$ and for all non-negative functions $\eta \in C_0^2(\mathbb{R})$. (The curve parameter t is an affine parameter.)

Let us indicate what QEIs and ANEC can say if they hold for the expected stress energy tensor $\langle \mathbf{T}_{ab}(x) \rangle_\omega$ which fulfills a semiclassical Einstein equation. It has been shown that the occurrence of spacetime geometries (M, g_{ab}) representing wormhole- or warp-drive scenarios as solutions to the semiclassical Einstein equations is very unlikely, if not impossible, provided that the expected stress energy fulfills QEIs (say, for the set of Hadamard states).[19] However, these arguments still rely on some approximations, and would really be completed once sufficiently sharp absolute QEIs are available. It seems that the work of Fewster and Smith[16] is close to providing that for the minimally coupled linear scalar field, but to our knowledge, the matter is not completely settled yet.

Another point can be made for ANEC. It has been shown that ANEC is sufficient to conclude singularity theorems for solutions to the semiclassical Einstein equations (which are similar to those for the classical Einstein equations in the presence of the weak energy condition).[2,34,41]

Thus, it appears that (suitable versions of) QEIs serve the purpose of ruling out solutions to the semiclassical Einstein equations which have pathological causal properties, and ANEC can be seen in a similar light, particularly as a property from which singularity theorems can be concluded. However, it is not clear if, e.g. for the minimally coupled scalar field on a globally hyperbolic spacetime, ANEC can be established for generic Hadamard states.

4. Local Thermal Equilibrium (LTE) States

We will now specialize our considerations with regard to averaged energy inequalities to a subset of the set of Hadamard states, called the set of *local thermal equilibrium states.* (Abbreviated, LTE states.) The concept of LTE states was introduced by Buchholz, Ojima and Roos.[7] The idea is best explained if one first considers a generic quantum field ϕ (taken for simplicity as a scalar field) on Minkowski spacetime fulfilling Wightman's axioms. Under very general conditions, such a quantum field will possess global thermal equlibrium states.[6] These states are distinguished by a rest frame and a time direction with respect to which they are in thermal equilibrium (i.e. are KMS-states), and an inverse temperature. These parameters can be gathered into a timelike, future-directed "inverse temperature vector" $\boldsymbol{\beta}^a$. At any given spacetime point x, there are certain observables which are pointlike concentrated at x and are sensitive to thermal properties when evaluated on the global thermal equilibrium states. For instance, for the massless linear scalar field ϕ_0, the Wick square $:\phi_0^2:(x)$ evaluated on a global thermal equilibrium state characterized by an inverse temperature vector $\boldsymbol{\beta}^a$ yields the value $c\boldsymbol{\beta}^a\boldsymbol{\beta}_a$, where c is a fixed constant. There are similar other pointlike observables which are, for a linear scalar field, typically of the form of (linear combinations of) Wick powers and their so-called "balanced derivatives" (see[7] for definition) and which correspond to intensive, density-like thermal quantities when evaluated on global equilibrium states. Let us collect a set S_x of such pointlike obsevables localized at x and sensitive to thermal properties. Fixing a spacetime point x, a state $\langle\,.\,\rangle_\omega$ of the quantum field is called an *LTE state* at x (with respect to S_x) if there is some inverse temperature vector $\boldsymbol{\beta}^a(x)$ with corresponding global thermal equilibrium state $\langle\,.\,\rangle_{[\boldsymbol{\beta}^a(x)]}$ so that $\langle\,.\,\rangle_\omega$ agrees with $\langle\,.\,\rangle_{[\boldsymbol{\beta}^a(x)]}$ on all observables in S_x,

$$\langle s_x\rangle_\omega = \langle s_x\rangle_{[\boldsymbol{\beta}^a(x)]} \tag{8}$$

for all $s_x \in S_x$. If N is a set of spacetime points and if $T > 0$, then one defines as $\mathcal{L}(N,T)$ the set of all states of the qantum field so that (8) holds for all $s_x \in S_x$, $x \in N$, with $[\boldsymbol{\beta}^a(x)\boldsymbol{\beta}_a(x)]^{-1} \leq T^2$. This corresponds then to the set of states which fulfill the condition of local thermal equilibrium at the spacetime points in the set N, and with a maximal bound T on the local temperatures which can be attained at the points in N. We refer to[5,7] for considerable further discussion of these concepts and examples.

These concepts can be carried over to the quantized linear scalar field ϕ on a globally hyperbolic spacetime (M, g_{ab}) (and also to more general quantum fields in the presence of local covariance, see[8]). The main modification is that $\langle\, . \,\rangle_{[\boldsymbol{\beta}^a(x)]}$ on the right hand side of (8) is now not interpreted as a state of the field ϕ on (M, g_{ab}), but as a state of the "Minkowski-space version" ϕ_{Min} of ϕ, identifying T_xM with Minkowski spacetime using geodesic normal coordinates around x. Then one needs also an identification between the S_x-observables of ϕ on (M, g_{ab}) and the corresponding S_x^{Min}-observables of ϕ_{Min} on Minkowski spacetime. For linear scalar quantum fields, this can be achieved by defining S_x as the set containing the Wick square $:\phi^2:(x)$ and its second balanced derivative $\eth_{ab} :\phi^2:(x)$. These quantities can be defined as expectation values on Hadamard states in a local covariant manner, and they have flat spacetime counterparts $:\phi^2_{Min}:(x)$ and $\eth_{ab} :\phi^2_{in}:(x)$. We shall not pause to explain this in any detail and refer to the more complete discussion in.[34] The basic contention is then — presented here with some simplifications, see again[34] for full explanations — to define a Hadamard state $\langle\, . \,\rangle_\omega$ of ϕ on (M, g_{ab}) as an LTE state at $x \in M$ if there is a global thermal equilibrium state $\langle\, . \,\rangle_{[\boldsymbol{\beta}^a(x)]}$ of ϕ_{Min} on Minkowski spacetime such that

$$\langle :\phi^2:(x)\rangle_\omega = \langle :\phi^2_{Min}:(x)\rangle_{[\boldsymbol{\beta}^a(x)]} \quad \text{and} \tag{9}$$

$$\langle \eth_{ab} :\phi^2:(x)\rangle_\omega = \langle \eth_{ab} :\phi^2_{Min}:(x)\rangle_{[\boldsymbol{\beta}^a(x)]} \,. \tag{10}$$

The point x on the right hand side of these equations is actually to be interpreted as the origin of Minkowski spacetime under an (x-dependent) identification of T_xM with Minkowski-spacetime using geodesic normal coordinates. Given a subset N of the spacetime M and some $T > 0$, one again collects in the set $\mathcal{L}(N, T)$ all Hadamard states $\langle\, . \,\rangle_\omega$ satisfying conditions (9) and (10) for all $x \in N$ with suitable $\boldsymbol{\beta}^a(x)$ so that $[\boldsymbol{\beta}^a(x)\boldsymbol{\beta}_a(x)]^{-1} \leq T^2$. (See, however,[34] for full details on this condition.)

It has then been shown in[34] that timelike and lightlike geodesically averaged QEIs hold for the states in $\mathcal{L}(N, T)$ as long as the local temperature T of the LTE states stays bounded. In other words, for LTE states of the linear scalar field on an arbitrary curved spacetime, one obtains an estimate of the form (6) for all states $\langle\, . \,\rangle_\omega \in \mathcal{L}(N, T)$ for any positive but fixed T and, of course, as long as the support of $h \circ \gamma$ is contained in N. We emphasize that this holds not only for timelike, but also for lightlike (or "null") geodesics, and for non-minimally coupled linear scalar quantum fields. The lower bounds $q(\gamma, h)$ depend, apart from γ and h, only on T and the parameters entering into the definition of the scalar quantum field model and depend in a local and covariant manner on the spacetime geometry. Thus, the QEIs for the LTE states of the linear scalar quantum field are also absolute QIEs.

Moreover, the validity of ANEC has been shown in[34] for LTE states of the quantized linear scalar scalar field on a curved spacetime. More precisely, if γ is a complete lightlike geodesic in a globally hyperbolic spacetime (M, g_{ab}), and if $\langle\,.\,\rangle_\omega$ is a Hadamard state on $\mathcal{F}(M, g_{ab})$, and if that state fulfills conditions (9) and (10) on all points x on the geodesic γ, then (7) holds for $\langle\,.\,\rangle_\omega$, for all values of curvature coupling $0 \leq \xi \leq 1/4$, provided that the constants renormalization constants A_1, A_2, A_3, A_3 take certain values, and provided that the "local temperature" $[\boldsymbol{\beta}^a(x)\boldsymbol{\beta}_a(x)]^{-1/2}$ grows moderately along the geodesic γ. The required conditions are given in detail in.[34]

5. Discussion

We have seen that a range of QEIs are now available for linear (scalar) quantum field models, and that many of them have the potential to rule out solutions to the semiclassical Einstein equations that exhibit pathological causal behaviour. However, to have good quantitative control about the scales down to which pathological spacetime geometries are really ruled out would require still better absolute QEIs with sharp bounds. One potential application would be to rule out the possibility that the apparent global equilibrium of the Universe at large scales could have come about by wormholes connecting spatially distant parts of the world shortly after the big bang, although this would need sharp absolute QEIs for interacting quantum fields. So far, little is known about QEIs for interacting quantum fields. Fewster and Hollands have obtained a nice general result on QEIs for conformal quantum field theories in 2-dimensional spacetime,[12] but the method of proof can't be generalized to higher dimensions. There are arguments that QEIs for interacting quantum fields should involve averaging of the expected energy density not over causal curves but over spacetime volumes.[10] However, the fact that QEIs are violated for the non-minimally linear scalar field, and that only relative QEIs can be established in this case,[13] lets one expect that relative QEIs are probably the best one can hope to obtain in general interacting quantum field theories. It is unknown if relative QEIs can be used to constrain pathological causal behaviour of spacetime geometries in solutions to the semiclassical Einstein equations.

The situation may improve for LTE states. There is good motivation to assume that LTE states are the states modelling matter states in early cosmology. However, it is very difficult to establish even the existence of LTE states in quantum field theory in curved spacetime (in fact, existence has not been proved to date in curved spacetime). It has been shown that there are LTE states which do not coincide with global thermal equlibrium states for the massless linear scalar field in Minkowski spacetime,[5] and similar results are also available for other types of linear quantum field models. The research on LTE states and their relation to QEIs therefore continues to be at its challenging initial stage, as was research on quantum field theory at the beginning of Wolfhard Zimmermann's career.

References

1. Alcubierre, M., "The Warp Drive: Hyper-Fast Travel within General Relativity", Class. Quant. Grav. **11** (1994) L73
2. Borde, A., "Geodesic Focussing, Energy Conditions and Singularities", Class. Quant. Grav. **4** (1987) 343
3. Brunetti, R., Fredenhagen, K., "Microlocal Analysis and Interacting Quantum Field Theories: Renormalization on Physical Backgrounds", Commun. Math. Phys. **208** (2000) 623
4. Brunetti, R., Fredenhagen, K., Verch, R., "The Generally Covariant Locality Principle – A New Paradigm for Local Quantum Field Theory", Commun. Math. Phys. **237** (2003) 31
5. Buchholz, D., "On Hot Bangs and the Arrow of Time in Relativistic Quantum Field Theory", Commun. Math. Phys. **237** (2003) 271
6. Buchholz, D., Junglas, P., "On The Existence Of Equilibrium States In Local Quantum Field Theory", Commun. Math. Phys. **121** (1989) 255
7. Buchholz, D., Ojima, I., Roos, H.-J., "Thermodynamical Properties of Non-Equilibrium States in Quantum Field Theory", Annals Phys. **297** (2002) 219
8. Buchholz, D., Schlemmer, J., "Local Temperature in Curved Spacetime", Class. Quantum Grav. **24** (2007) F25
9. Epstein, H., Glaser, V., Jaffe, A., "Nonpositivity of the Energy Density in Quantized Field Theories", Nuovo Cimento **36** (1965) 1016
10. Fewster, C.J., "Energy Inequalities in Quantum Field Theory", Expanded and updated version of a contribution to the proceedings of the XIV ICMP, Lisbon 2003, arXiv:math-ph/0501073
11. Fewster, C.J, "A General Worldline Quantum Inequality", Class. Quant. Grav. **17** (2000) 1897; Fewster, C.F., Verch, R., "A Quantum Weak Energy Inequality for Dirac Fields in Curved Spacetime", Commun. Math. Phys. **225** (2002) 331; Fewster, C.J., Pfenning, M.J., "A Quantum Weak Energy Inequality for Spin-One Fields in Curved Spacetime", J. Math. Phys. **44** (2003) 4480
12. Fewster, C.J., Hollands, S., "Quantum Inequalities in Two-Dimensional Conformal Field Theory", Rev. Math. Phys. **17** (2005) 577

13. Fewster, C.J., Osterbrink, M., "Quantum Energy Inequalities for the Non-Minimally Coupled Scalar Field", J. Phys. **A 41** (2008) 025402
14. Fewster, C.J., Pfenning, M.J., "Quantum Energy Inequalities and Local Covariance I: Globally Hyperbolic Spacetimes", J. Math. Phys. **47** (2006) 082303; Fewster, C.J., "Quantum Energy Inequalities and Local Covariance II: Categorical Formulation", Gen. Rel. Grav. **39** (2007) 1855
15. Fewster, C.J., Roman, T.A., "Null energy conditions in Quantum Field Theory", Phys. Rev. **D 67** (2003) 044003
16. Fewster, C.J., Smith, C.J., "Absolute Quantum Energy Inequalities in Curved Spacetime", Ann. H. Poincaré, to appear, arXiv:gr-qc/0702056
17. Ford, L.H., "Quantum Coherence Effects and the Second Law of Thermodynamics", Proc. R. Soc. Lond. A **364** (1978) 227
18. Ford, L.H., "Constraints on Negative Energy Fluxes", Phys. Rev. **D 43** (1991) 3972; Ford, L.H., Roman, T.A., "Averaged Energy Conditions and Quantum Inequalities", Phys. Rev. **D 51** (1995) 4277; — "Restrictions on Negative Energy Density in Flat Space-Time", Phys. Rev. **D 55** (1997) 2082; — "The Quantum Interest Conjecture", Phys. Rev. **D 60** (1999) 104018
19. Ford, L.H. Roman, T.A., "Quantum Field Theory Constrains Traversable Wormhole Geometries", Phys. Rev. **D 53** (1996) 5496-5507; Ford, L.H., M.J. Pfenning, "The Unphysical Nature of 'Warp Drive'", Class. Quant. Grav. **14** (1997) 1743; Fewster, C.J., Roman, T.A., "On Wormholes with Arbitrarily Small Quantities of Exotic Matter", Phys. Rev. **D 72** (2005) 044023
20. Haag, R., Kastler, D., "An Algebraic Approach to Quantum Field Theory", J. Math. Phys. **5** (1964) 848
21. Hawking, S.W., "Particle Creation by Black Holes", Commun. Math. Phys. **43** (1975) 199; Wald, R.M., "On Particle Creation by Black Holes", Commun. Math. Phys. **45** (1975) 9; Fredenhagen, K., Haag, R., "On The Derivation of Hawking Radiation Associated with the Formation of a Black Hole", Commun. Math. Phys. **127** (1990) 273
22. Hawking, S.W., Ellis, G.F.R., *The Large Scale Structure of Space-Time*, CUP, Cambridge, 1973
23. Hollands, S., "Renormalized Quantum Yang-Mills Fields in Curved Spacetime", arXiv:0705.3340
24. Hollands, S., Wald, R.M., "Local Wick Polynomials and Time Ordered Products of Quantum Fields in Curved Spacetime", Commun. Math. Phys. **223** (2001) 289; — "Existence of Local Covariant Time Ordered Products of Quantum Fields in Curved Spacetime", Commun. Math. Phys. **231** (2002) 309
25. Hollands, S., Wald, R.M., "Axiomatic Quantum Field Theory in Curved Spacetime", arXiv:0803.2003
26. Jost, R., *The Generalized Theory of Quantum Fields*, Amer. Math. Soc., Providence, Rhode Island, 1965
27. Kay, B.S., Wald, R.M., "Theorems on the Uniqueness and Thermal Properties of Stationary, Nonsingular, Quasifree States on Space-Times with a Bifurcate Killing Horizon", Phys. Rep. **207** (1991) 49
28. Kriele, M., *Spacetime*, Springer LNP m59, Springer-Verlag, Berlin, 2001

29. Lehmann, H., Symanzik, K., Zimmermann, W., "Zur Formulierung quantisierter Feldtheorien", Nuovo Cimento **1** (1955) 205; — "On the Formulation of Quantized Field Theories. II", Nuovo Cimento **6** (1957) 319
30. Morris, M.S., Thorne, K., Yurtsever, U., "Wormholes, Time Machines, and the Weak Energy Condition", Phys. Rev. Lett. **61** (1988) 1446
31. Moretti, V., "Comments on the Stress-Energy Tensor Operator in Curved Spacetime", Commun. Math. Phys. **232** (2003) 189
32. Parker, L., Fulling, S.A., "Quantized Matter Fields and the Avoidance of Singularities in General Relativity", Phys. Rev. **D 7** (1973) 2357
33. Radzikowski, M.J., "Micro-Local Approach to the Hadamard Condition in Quantum Field Theory in Curved Spacetime", Commun. Math. Phys. **179** (1996) 529
34. Schlemmer, J., Verch, R., "Local Thermal Equilibrium States and Quantum Energy Inequalities", Ann. H. Poincaré, to appear, arXiv:0802.2151
35. Streater, R.F., Wightman, A.S., *PCT, Spin and Statistics, and All That*, 2nd ed., Benjamin, New York, 1968
36. Strohmaier, A., Verch, R., Wollenberg, M., "Microlocal Analysis of Quantum Fields on Curved Spacetimes: Analytic Wave-Front Sets and Reeh-Schlieder Theorems", J. Math. Phys. **43** (2002) 5514
37. Verch, R., "A Spin-Statistics Theorem for Quantum Fields on Curved Spacetime Manifolds in a Generally Covariant Framework", Commun. Math. Phys. **223** (2001) 262
38. Wald, R.M., "Trace Anomaly of a Conformally Invariant Quantum Field in Curved Spacetime", Phys. Rev. **D17** (1978) 1477; — "The Back Reaction Effect in Particle Creation in Curved Space-Time", Commun. Math. Phys. **54** (1977) 1
39. Wald, R.M, *General Relativity*, University of Chicago Press, Chicago, 1984
40. Wald, R.M., *Quantum Field Theory in Curved Spacetime and Black Hole Thermodynamics*, University of Chicago Press, Chicago, 1994
41. Wald, R.M., Yurtsever, U., "General Proof of the Averaged Null Energy Condition for a Massless Scalar Field in Two-dimensional Curved Spacetime", Phys. Rev. **D 44** (1991) 403

FIELD THEORY AND BRANE DYNAMICS

THOMAS E. CLARK*

Department of Physics, Purdue University
West Lafayette, IN 47907-2036, USA
**E-mail: clark@physics.purdue.edu*

The formation of a brane in the bulk spontaneously breaks its space-time symmetries down to that of the world volume and its complement. The long wavelength oscillations of the brane world are described by massive brane vector gauge fields. The coupling of the brane vectors to Standard Model fields is determined by the method of nonlinear realizations of broken space-time symmetries of the bulk. Data from LEP and the Tevatron are used to exclude regions of brane vector parameter space while the regions accessible to the LHC are presented.

Keywords: Extra dimensions; Brane vector; Brane vector collider production; Nonlinear realizations and brane oscillations

In Celebration Of Wolfhart Zimmermann's 80th Birthday.

1. Introduction

When a brane,[1] such as a domain wall, is present, field theory requires modification, at least in part, due to the spontaneous breaking of the space-time symmetries of the bulk to those of the world volume of the embedded brane and its complement. At low energy, the spectrum of particles must contain the Nambu-Goldstone boson describing the oscillation of the brane into the co-volume.[2] Since the gravitational field is dynamical, the Nambu-Goldstone boson will be eaten by the zero mode graviphoton, making it a world volume massive (Proca) vector field.[3] In the case of compact extra dimensions and for a flexible brane, one whose tension is less than the Planck scale of the bulk, this oscillation will couple more strongly to the Standard Model particles than the higher Kaluza-Klein modes of the bulk fields.[4] These massive brane vector particles can couple to the Standard Model particles through their energy-momentum tensor. In addition, the extrinsic curvature of the brane provides a new means of coupling the Standard Model to the brane vectors. These interactions offer a new source for detecting the existence of extra space dimensions. The brane vectors can be produced in colliders and escape detection. As such, they appear as missing energy. The signature of the production of a single photon and missing energy at LEP, the Tevatron and in the future at the LHC provides excluded and allowed regions of brane vector parameter space as well as a means of discovery of brane oscillations.[5,6] These constraints and expectations will be reviewed in this article. Beyond the scope of the present work is a discussion of brane vectors as dark matter. The relic abundance and direct detection experiments are an additional source of discovery or constraints to the brane vector parameter space.[7] In addition, the coupling of the brane vector to the Higgs field offers a new invisible Higgs decay channel.[8]

The next section of this Festschrift contribution deals with the low energy effective action of a flexible brane.[2] The long wavelength degrees of freedom localized on the brane are described by massive vector (Proca) fields.[3] The form of the leading interactions with the assumed brane localized Standard Model fields is determined by means of nonlinearly realizing the broken Poincaré symmetries of the bulk due to the embedded brane. The interaction is characterized by the effective brane tension, the mass of the brane vector and the strength of the coupling to the Standard Model energy-momentum tensor and the extrinsic curvature of the brane. In section 3 the data from LEP and the Tevatron are used to delineate excluded and allowed regions of the brane vector parameter space. As well, regions of parameter space that will be accessible or inaccesible to the LHC are presented.

Before beginning I would like to wish Wolfhart a very Happy Birthday! At times like this one tends to reminise. The energy-momentum tensor has played a central role in much of the physics I have done over the years, with the current work being no exception. Of course, it was Wolfhart that set me on this much appreciated path. As his graduate student he suggested that I use the then just developed BPHZ program to construct a finite energy-momentum tensor in A^4 theory. This was a great problem. It involved his proof of the short distance expansion and in general, the importance and treatment of symmetries in renormalized perturbation theory. I fondly recall the many evening sessions at the Courant Institute when Wolfhart spent much time with this naive student's tedious calculations. The seemingly impromtu discussions on scaling symmetry, point splitting, Riesz's theorem and so many other topics opened the world of theoretical physics for me. He was always supportive of my career and deserves a sincere thank you. He invited me to München for post doctoral work. He suggested I work on some specific issues in gauge theory and what great experience that topic provided, especially as I learned to walk on my own. The exposure to great physics and physicists was and is fantastic at the Max Planck Institute. While there, I met Klaus Sibold and Olivier Piguet and we began an enjoyable collaboration working on renormalization of supersymmetric theories. The renormalization of the supercurrent was a central project. Of course, one of the components of the supercurrent is the good old, or should I say, new impoved energy-momentum tensor.

As my career progressed through various post doctoral assignments I maintained contact with Wolfhart during this time. These assignments led me to my current position at Purdue University. There I met my long time collaborator and friend Sherwin Love. We have frequently turned to symmetries if not the energy-momentum tensor per se as a source of fruitful research. Always interested and encouraging, Wolfhart spent a sabbatical semester at Purdue in 1986, staying in a house provided by Sherwin while he was on his sabbatical. Wolfhart and my wife, Nancy, a music educator, enjoyed discussing music and instruments especially in light of the musical talent of Wolfhart's children (embouchure is what?).

Physics research continues and it is a pleasure to have met and worked with so many talented people. None more so than my colleagues that worked on the topics to be discussed here. Tonnis ter Veldhuis, who is my former student and with whom I have a special bond, now a professor at Macalester College in St. Paul, Minnesota, Muneto Nitta is a professor at Keio University in Tokyo and is a former post doctoral research associate at Purdue University and naturally Sherwin Love who is my colleague at Purdue. As will be seen, the energy-momentum tensor is playing a vital role in brane world physics....... It seems but a short time ago since the halcyon days at NYU.

Many more Wolfhart!

2. Massive Brane Vectors

The formation of a brane in the bulk spontaneously breaks the higher dimensional space-time symmetries down to those of the world volume and its complement. The concommitant brane localized Nambu-Goldstone bosons describe the oscillations of the brane into its covolume. The nonlinear realization of the broken symmetries by the coset method provides the covariant building blocks to construct an action that is invariant under the larger bulk Poincaré symmetries.[2] Assuming that the Standard Model fields are localized on the brane, the invariant form of their interaction with the brane oscillations is also prescribed by this method. Consider initially the case of a domain wall embedded in $D = 5$ dimensional space-time. Grouping the bulk Poincaré generators according to their representation under the unbroken $SO(1,3)$ world volume Lorentz transformations, $P^M = (P^\mu, Z = -P^{M=4})$ and $M^{MN} = (M^{\mu\nu}, K^\mu = 2M^{4\mu})$. For the case of broken space-time symmetries, the coset element $U(x) \in ISO(1,4)/SO(1,3)$

$$U(x) = e^{ix^\mu P_\mu} e^{i\phi(x)Z} e^{iv^\mu(x)K_\mu}. \tag{1}$$

The coordinates that parameterize the coset not only include the Nambu-Goldstone boson brane oscillation field $\phi(x)$ associated with the broken translation generator denoted Z but also the world volume $D = 4$ dimensional space-time coordinates x^μ associated with the unbroken world volume space-time translation generators P^μ. In addition the coset includes the auxiliary Nambu-Goldstone boson fields $v^\mu(x)$ associated with the broken Lorentz transformation generators K_μ.

The bulk Poincaré algebra valued Maurer-Cartan form, $U^{-1}(x)\partial_\mu U(x)$, supplies the covariant building blocks for the action

$$U^{-1}(x)\partial_\mu U(x) = i \left[e_\mu{}^m(x) P_m + \nabla_\mu \phi(x) Z + \nabla_\mu v^m(x) K_m + \omega_\mu^{mn}(x) M_{mn} \right]. \tag{2}$$

The brane oscillations induce a vierbein $e_\mu{}^m(\phi, v^m)$ and spin connection $\omega_\mu^{mn} = v^m \partial_\mu v^n - v^n \partial_\mu v^m + \cdots$ on the world volume. The covariant derivative of the Nambu-Goldstone field, $\nabla_\mu \phi(x)$, can be covariantly constrained to zero, $\nabla_\mu \phi = 0$, thus allowing the elimination of the auxiliary field $v_\mu = \partial_\mu \phi + \cdots$ through the "inverse Higgs mechanism".[9] With this constraint the vierbein becomes

$$e_\mu{}^m = \delta_\mu{}^m - \frac{\partial_\mu \phi \partial^m \phi}{(\partial\phi)^2} \left(\frac{1 - \sqrt{1 - (\partial\phi)^2}}{(\partial\phi)^2} \right), \tag{3}$$

and the determinant of the vierbein reduces to $\det e = \sqrt{1 - \partial_\mu \phi \partial^\mu \phi}$. As well, the covariant derivative of the auxiliary Nambu-Goldstone field, $\nabla_\mu v^m$, is related to the extrinsic curvature of the embedded brane which is given by $K_{\mu\nu} = e_\nu{}^m \nabla_\mu v_m = \partial_\mu v_\nu + \cdots = \partial_\mu \partial_\nu \phi + \cdots$ and describes the rigidity or stiffness of the brane.[10]

The $ISO(1,4)$ invariant action is just the Nambu-Goto action of the brane oscillations[11] and the induced gravity action of the Standard Model fields[12]

$$\Gamma_{\text{N-G}} = \int d^4x \det e[-f^4 + \mathcal{L}_{\text{SM}}(e)], \tag{4}$$

with the brane tension given by $\sigma = f^4$ and $\mathcal{L}_{\rm SM}(e)$ the Standard Model Lagrangian coupled to the brane oscillations through the induced vierbein. Expanding the action in terms of derivatives of the brane oscillation fields for small oscillations of the brane relative to the tension and adding a mass m for the scalar, the leading order interactions are obtained

$$\Gamma_{\phi {\rm SM}} = \int d^4x \left[\frac{1}{2f^4} \partial^\mu \phi \partial^\nu \phi T^{\rm SM}_{\mu\nu} - \frac{m^2}{8f^4} \phi^2 \eta_{\mu\nu} T^{\rm SM}_{\mu\nu} \right], \tag{5}$$

where $T^{\rm SM}_{\mu\nu}$ is the Standard Model energy-momentum tensor. The subsequent phenomenology for massless[13] as well as massive[14] brane oscillation scalars has been throughly studied.

The bulk gravitational fields however are dynamic. Consequently, the brane oscillation Nambu-Goldstone boson is eaten by the zero mode graviphoton field. Thus, the brane oscillation is characterized by a massive vector (Proca) field, denoted X_μ, localized on the world volume. Extending the coset method to include gravitational interactions[3] provides the effective interaction of the brane vectors with Standard Model fields. The missing energy production process of $f\ \bar{f} \to \gamma + XX \to \gamma + \not{E}$ with f a Standard Model fermion and $\bar{f}$ its antiparticle will provide bounds on the parameter space of the brane oscillations.

In the domain wall case, the $ISO(1,4)$ local symmetries of the bulk can be realized non-linearly by introducing the brane localized gravitational fields in a Poincaré algebra valued one-form[3]

$$E_\mu(x) = E_\mu{}^m(x) P_m + X_\mu(x) Z + V^m_\mu(x) K_m + \frac{1}{2}\gamma^{mn}_\mu(x) M_{mn}, \tag{6}$$

with $E_\mu{}^m(x)$ the dynamic gravitational vierbein on the brane and $\gamma^{mn}_\mu(x)$ the related spin connection. The brane vector field is associated with the broken bulk space translation generator Z and in this codimension $N = 1$ case is just the single field $X_\mu(x)$. The field associated with the broken Lorentz transformations, $V^m_\mu(x)$, will enter the effective action as an auxiliary field related to the second covariant derivative of ϕ which is already one of the covariant building blocks of the action. The locally covariant Maurer-Cartan form is defined as

$$\begin{aligned} &U^{-1}(x) \left[\partial_\mu + i e^{ix\cdot P} E_\mu(x) e^{-ix\cdot P} \right] U(x) \\ &\qquad = i \left[e_\mu{}^m P_m + \nabla_\mu \phi Z + K^m{}_\mu K_m + \omega^{mn}_\mu M_{mn} \right], \end{aligned} \tag{7}$$

where now the component one-forms depend on the gravitational fields as well as the Nambu-Goldstone fields.

The broken local Lorentz transformations can be used to transform to a partial unitary gauge in which the auxiliary Nambu-Goldstone field $v_\mu(x) = 0$. In this gauge the components of the Maurer-Cartan form become $e_\mu^{\ m}(x) = \delta_\mu^{\ m} + E_\mu^{\ m}(x) + 2\phi(x)V_\mu^m(x)$ and the covariant derivative $\nabla_\mu\phi(x) = \partial_\mu\phi(x) + X_\mu(x)$. The spin connection is just the gravitational term, $\omega_\mu^{mn}(x) = \gamma_\mu^{mn}(x)$ as is the covariant derivative of v^m, $\nabla_\mu v^m(x) = V_\mu^m(x)$. The coset method guarantees the bulk invariance of the action as long as it is world volume invariant and constructed using these Maurer-Cartan building blocks. The transformation of the Standard Model fields can be extended and similarly included in the invariant action building procedure. The terms involving the brane vector are made from covariant derivatives of the Nambu-Goldstone boson ϕ. The covariant field strength tensor for the brane vector is obtained from the commutator of covariant derivatives, $F_{\mu\nu} = [\nabla_\mu, \nabla_\nu]\phi = \partial_\mu X_\nu - \partial_\nu X_\mu$. The anti-commutator of covariant derivatives begins with $\partial_\mu\partial_\nu\phi$ and as such will be referred to as the extrinsic curvature[15] as only its leading terms will be needed in the expansion of the effective action, $K_{\mu\nu} \equiv 1/2\{\nabla_\mu, \nabla_\nu\}\phi$. These covariant building blocks are used to construct the locally invariant action which, after rescaling the fields to give them canonical dimension, has the form of a general coordinate invariant world volume action[3]

$$\Gamma = \int d^4x \det e \left[\Lambda + \frac{1}{2\kappa^2}R - \frac{1}{4}F_{\mu\nu}F^{\mu\nu} + \frac{1}{2}\nabla^\mu\phi\nabla_\mu\phi + \mathcal{L}_{\rm SM}(e) \right. \\ \left. + \frac{\tau}{F_X^4}\nabla^\mu\phi\nabla^\nu\phi T_{\mu\nu}^{\rm SM} + \frac{M_X^2}{2F_X^4}\left(K_1 B_{\mu\nu} + K_2\tilde{B}_{\mu\nu}\right)F^{\mu\rho}K_\rho^{\ \nu} + \cdots\right], \quad (8)$$

where F_X is the effective brane tension scale while M_X is the brane vector mass and τ, K_1 and K_2 are coupling constants. The ellipses denote additional terms such as a coupling to the invariant Higgs composite $\Phi^\dagger\Phi$, which can lead to invisible Higgs decay, as well as higher dimensional monomials.

Finally, the broken local space translation invariance can be used to transform to a full unitary gauge in which the Nambu-Goldstone scalar field $\phi(x) = 0$. The world volume vierbein then becomes purely gravitational, $e_\mu^{\ m}(x) = \delta_\mu^{\ m} + E_\mu^{\ m}(x)$ and the covariant derivative of the scalar is simply the mass of the brane vector times the field, $\nabla_\mu\phi(x) = M_X X_\mu(x)$ with the extrinsic curvature related term becoming $K_{\mu\nu} = 1/2(\partial_\mu X_\nu + \partial_\nu X_\mu) + \cdots$. Ignoring the purely world volume gravitational interactions and generalizing to the case of co-dimension $= N$ so that there are N-species of brane vector, X_μ^i, with $i = 1, 2, \ldots, N$, the effective action describing the interaction of the brane vector with the Standard Model fields is obtained[5,6]

$$\Gamma_{\text{Effective}} = \int d^4x \left[\mathcal{L}_{\text{SM}}(\eta) - \frac{1}{4} F^i_{\mu\nu} F^{\mu\nu}_i + \frac{1}{2} M_X^2 X^i_\mu X^\mu_i + \frac{\tau}{2} \frac{M_X^2}{F_X^4} X^\mu_i T^{\text{SM}}_{\mu\nu} X^\nu_i \right.$$
$$\left. + \frac{M_X^2}{2F_X^4} \left(K_1 B_{\mu\nu} + K_2 \tilde{B}_{\mu\nu} \right) F^{i\mu\rho} K^{i\nu}_\rho \right], \tag{9}$$

where only bilinear terms in the brane vector fields are considered in $\Gamma_{\text{Effective}}$ so that $K^i_{\mu\nu} = 1/2(\partial_\mu X^i_\nu + \partial_\nu X^i_\mu)$. The covolume $SO(N)$ symmetry is envisioned to be spontaneously broken, hence, the associated gauge fields are massive and not considered here. Although the $SO(N)$ symmetry amongst the brane vectors is now broken, the brane vectors are taken to have the same mass M_X and effective brane tension F_X. Similarly, the bilinear X coupling can be to any $SU(3) \times SU(2) \times U(1)$ invariant. These have been chosen to be equal, hence the Standard Model energy-momentum tensor appears with overall coupling strength τ.

3. Collider Bounds On Brane Vector Parameter Space

The annihilation of a fermion f and anti-fermion $\bar{f}$ to produce a single photon and a pair of brane vectors which escape detection and appear as missing energy provides a means to bound the brane vector parameter space. Data from the LEP-II process $e^+ e^- \to \gamma + XX \to \gamma + \not{E}$ and the Tevatron-II process $p\,\bar{p} \to \gamma + XX \to \gamma + \not{E}$ are in agreement with the Standard Model for the production of single photon plus missing energy events. This puts a bound on the production cross section calculated from the above effective action and hence determines an allowed and excluded region of the F_X, M_X, N, τ, K_1 and K_2 parameter space. The Feynman diagrams contributing to the production process are shown in Fig. 1. The differential cross section for spin averaged incoming fermion and anti-fermion collisions producing a photon and 2 brane vectors, summed over all polarizations and X species $i = 1, 2, \ldots, N$ is given by[6]

$$\frac{d^2\sigma_{\gamma XX}}{dk^2 dt} = \frac{\alpha}{4\pi} \frac{1}{15,360\pi} \left[\frac{N}{F_X^8} \right] \frac{1}{\hat{s}^3 ut} \frac{\sqrt{k^2 - 4M_X^2}}{\sqrt{k^2}} \left[2\hat{s}k^2 + u^2 + t^2 \right]$$
$$\times \left\{ \tau^2 \left(\hat{s}k^2 + 4ut \right) \left(\left[k^2 - 4M_X^2 \right]^2 + 20M_X^2 \left(k^2 + 2M_X^2 \right) \right) \right.$$
$$\left. + \left[K_1^2 k^2 + K_2^2 \left(k^2 - 4M_X^2 \right) \right] SM \left[80\hat{s} \left(k^2 \right)^2 M_X^2 \right] \right\}, \tag{10}$$

where α is the electromagnetic fine structure constant evaluated at the W mass. SM is the Standard Model factor coming from the photon or Z

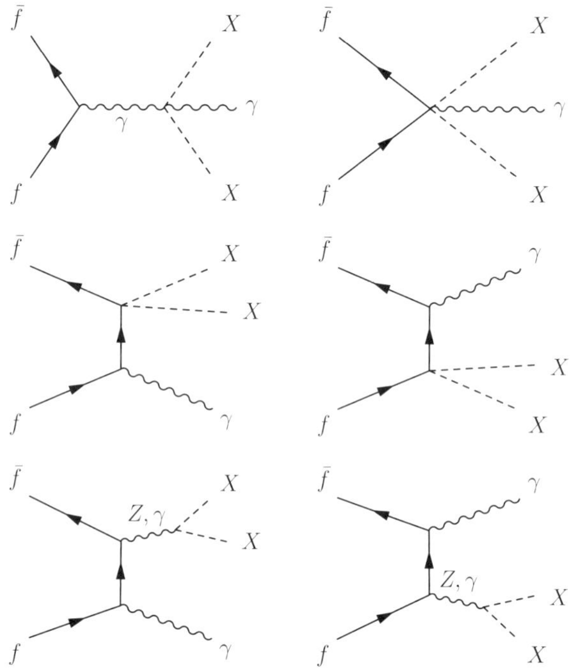

Fig. 1. The Feynman graphs contributing to the single photon and missing energy brane vector production. The top four graphs involve $T^{\rm SM}_{\mu\nu}$ while the bottom two graphs depend on the coupling to $K^i_{\mu\nu}$.

exchange in the last 2 graphs in Fig. 1

$$SM = \pi\alpha \left\{ \frac{\cos^2\theta_W}{(k^2)^2} + \frac{1}{(k^2 - M_Z^2)^2 + M_Z^2\Gamma_Z^2} \right. \times \left[\frac{1}{\cos^2\theta_W}\left(\frac{1}{16} + \left(\sin^2\theta_W - \frac{1}{4}\right)^2\right) \right. \left. \left. + 2\cos^2\theta_W \left(\sin^2\theta_W - \frac{1}{4}\right)\frac{(k^2 - M_Z^2)}{k^2}\right]\right\}, \tag{11}$$

with θ_W the Weinberg angle and M_Z and Γ_Z the mass and width of the Z gauge boson. In particular there is no interference between the energy-momentum tensor coupling terms τ and the extrinsic curvature related interactions K_1 and K_2.

The anti-fermion 4-momentum is p_1 and the fermion 4-momentum p_2. The photon momentum is q with the 2 brane vectors' 4-momenta k_1 and k_2. The Mandelstam variables $\hat{s}$, t, u and k^2 are introduced. In terms of the particles' momenta the variables are given by $\hat{s} = (p_1 + p_2)^2$, $t = (p_1 - q)^2$, $u = (p_2 - q)^2$ and $k^2 = (k_1 + k_2)^2$. Since the fermion masses are neglected $\hat{s} + t + u = k^2 + q^2 + p_1^2 + p_2^2 = k^2$. For an $e^+ \ e^-$ collider $\hat{s} = s$ and the positron's and electron's energies are half the center of mass energy $\sqrt{s}$, that is the beam energy, $E_{\text{Beam}} = \sqrt{s}/2$. For the Tevatron collider, the proton has the beam energy while the subprocess (anti-)quark carries the fraction x of the proton energy, $x\sqrt{s}/2$, and the anti-proton also has half the center of mass energy while the subprocess (anti-)quark carries the fraction y of the anti-protron energy, $y\sqrt{s}/2$. Similarly for the LHC, each proton has half the center of mass energy while one subprocess (anti-)quark carries fraction x of that energy while the other subprocess (anti-)quark carries fraction y of its proton's energy. It follows that for the hadron colliders $\hat{s}$ is related to the center of mass energy by $\hat{s} = xys$.

The total cross section is found for the lepton collider by directly integrating over the kinematically allowed region of photon energy and angle with appropriate cuts. For the lepton collider it is the electron and positron that collide and so the integral is evaluated at the machine's center of mass energy $\sqrt{s} = 206$ GeV. For the hadron colliders the quark and anti-quark subprocess annihilation produces the photon and 2 brane vectors. The quark and anti-quark carry only fractions of the beam energy and so the differential cross section must be integrated with the quark distribution functions over the range of energies of the quarks, that is the x and y energy fractions over their kinematically allowed regions. The transverse energy of the photon is $E_{\gamma\text{T}} = E_\gamma \sin\theta$ where the photon's total energy is denoted E_γ. The photon's beam axis polar angle θ is expressed in terms of the pseudorapidity $\eta = -\ln\tan(\theta/2)$. In general the cuts on the polar angle of the photon are related to a minimum and maximum pseudorapidity. This then determines minimum and maximum t integration limits: $t_{\substack{\min\\ \max}} = (k^2 - \hat{s})x[1 - \tanh\eta_{\substack{\min\\ \max}}]/[(y+x) + (y-x)\tanh\eta_{\substack{\min\\ \max}}]$. The minimum cut on the transverse energy of the photon, $E_{\gamma\text{T}}^{\min}$, yields a maximum for the k^2 variable, $k^2_{\max} = \hat{s}(1 - 2E_{\gamma\text{T}}^{\min}/\sqrt{\hat{s}})$, while the minimum integration limit to produce 2 X particles is $k^2_{\min} = 4M_X^2$. This value is used for the hadronic cases. For LEP the transverse energy of the photon not only is bounded below but also is cut above.

For the lepton case, $\hat{s} = s$ and the total cross section is given by

$$\sigma_{\gamma XX}^{\text{LEP}} = \int_{k^2_{\min}}^{k^2_{\max}} dk^2 \int_{t_{\min}}^{t_{\max}} dt \frac{d^2\sigma_{\gamma XX}}{dk^2 dt}, \tag{12}$$

with the photon polar angle bounded by $|\cos\theta| \leq 0.97$ and the transverse photon energy in the range $0.04E_{\rm Beam} \leq E_{\gamma {\rm T}} \leq 0.6E_{\rm Beam}$.[16] For the Tevatron and the LHC the parton distribution functions must be folded in with the fractional energy integration over x and y where $\hat{s} = xys$. The lower integration limits for the energy fractions are $x_{\rm min} = \hat{s}_{\rm min}/s$ and $y_{\rm min} = x_{\rm min}/x$ with $\hat{s}_{\rm min} = 2E_{\gamma {\rm T}}^{\rm min 2} + 4M_X^2 + 2E_{\gamma {\rm T}}^{\rm min}\sqrt{E_{\gamma {\rm T}}^{\rm min 2} + 4M_X^2}$. Using the CTEQ-6.5M quark distribution functions[17] appropriate for the energy range and process, the total cross section for the hadron colliders takes the form

$$\sigma_{\gamma XX}^{\rm Hadron} = \int_{x_{\rm min}}^{1} dx \int_{y_{\rm min}}^{1} dy f(x, y; \hat{s}) \int_{k_{\rm min}^2}^{k_{\rm max}^2} dk^2 \int_{t_{\rm min}}^{t_{\rm max}} dt \frac{d^2\sigma_{\gamma XX}}{dk^2 dt}. \tag{13}$$

The quark distribution function appropriate for the $p\bar{p}$ collisions of the Tevatron is

$$\begin{aligned} f(x, y; \hat{s}) &= \frac{1}{3}\left(\frac{2}{3}\right)^2 \left[u_p(x, \hat{s})\bar{u}_{\bar{p}}(y, \hat{s}) + \bar{u}_p(x, \hat{s})u_{\bar{p}}(y, \hat{s})\right] \\ &\quad + \frac{1}{3}\left(-\frac{1}{3}\right)^2 \left[d_p(x, \hat{s})\bar{d}_{\bar{p}}(y, \hat{s}) + \bar{d}_p(x, \hat{s})d_{\bar{p}}(y, \hat{s})\right]. \end{aligned} \tag{14}$$

The quark distribution function appropriate for the pp collisions of the LHC is

$$\begin{aligned} f(x, y; \hat{s}) &= \frac{1}{3}\left(\frac{2}{3}\right)^2 \left[u_p(x, \hat{s})\bar{u}_p(y, \hat{s}) + \bar{u}_p(x, \hat{s})u_p(y, \hat{s})\right] \\ &\quad + \frac{1}{3}\left(-\frac{1}{3}\right)^2 \left[d_p(x, \hat{s})\bar{d}_p(y, \hat{s}) + \bar{d}_p(x, \hat{s})d_p(y, \hat{s})\right]. \end{aligned} \tag{15}$$

u_p denotes the fraction of up quark in the proton, and so on. The overall $1/3$ is the probability the quarks have the same color and the distributions f include the electric charge coupling of the quarks in units of e. The pseudorapidity is taken in the region $|\eta| \leq 1.0$ for the hadronic cases and the minimum transverse energy of the photon is $E_{\gamma {\rm T}}^{\rm min} = 45$ GeV for the Tevatron[18] and is scaled to $E_{\gamma {\rm T}}^{\rm min} = 350$ GeV for the LHC.

The number of single photon plus missing energy events at LEP and the Tevatron is in agreement with the Standard Model. Thus the number of events coming from the brane vector production cross section must be less than 5 standard deviations of the Standard Model background, $\sigma_{\gamma XX}\mathcal{L} \leq \sigma_{\rm Discovery}\mathcal{L} = 5\sqrt{N_{\rm SMbkgrnd}}$, with $\mathcal{L}$ the integrated luminosity for the collider data and $\sigma_{\rm Discovery}$ the maximum for the brane vector single photon cross section. This will provide excluded and allowed regions of brane vector parameter space. In the case of the LHC, the discovery cross section is estimated by just the increase in luminosity so that

$\sigma_{\text{Discovery}}(LHC) = \sqrt{\mathcal{L}_{\text{TeV-II}}/\mathcal{L}_{\text{LHC}}}\ \sigma_{\text{Discovery}}(TeVII)$. This will provide LHC accessible and inaccessible regions of brane vector parameter space. The parameter space bounds are illustrated in Fig. 2 for LEP-II, Figs. 3 and 4 for Tevatron-II and in Fig. 5 for the LHC reach. The line of exclusion/accessibility has a mild dependence, $\sim N^{1/8}$, on the number of extra space dimensions N, hence it is plotted for $N = 1$. The regions of parameter space shown in the Figs. 2, 3 and 5 are for 3 cases of fixed values of the coupling to the energy-momentum tensor, which is given by the value of τ, and the couplings to the extrinsic curvature which are given by the K_1 and K_2 values. Table 1 summarizes the collider properties and backgrounds.[16,18] The dependence of the cross section on the extrinsic curvature coupling constants can be exhibited for fixed effective brane tension F_X and brane vector mass M_X as plotted in Fig. 4 for the Tevatron for 2 sets of values.

Collider	$\sqrt{s}$ (TeV)	$\mathcal{L}$ (pb^{-1})	$\sigma_{\text{Discovery}}$ (pb)
LEP-II	0.206	138.8	0.45
Tevatron-II	1.96	84	0.25
LHC	14	10^5	0.0071

Table 1. The colliders, their center of mass energies, integrated luminosities and discovery cross-sections.

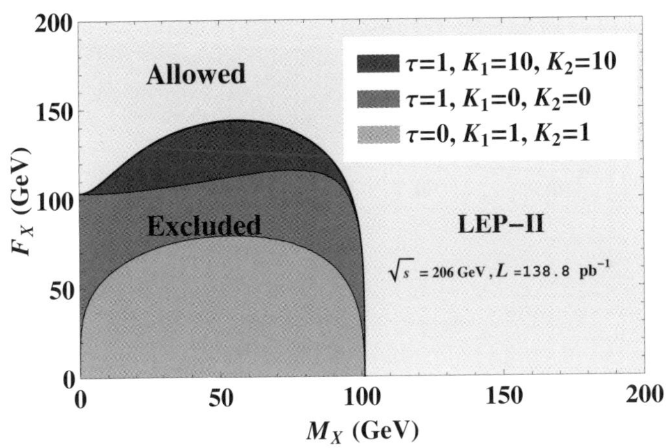

Fig. 2. LEP-II excluded regions of brane vector parameter space for fixed values of τ, K_1 and K_2.

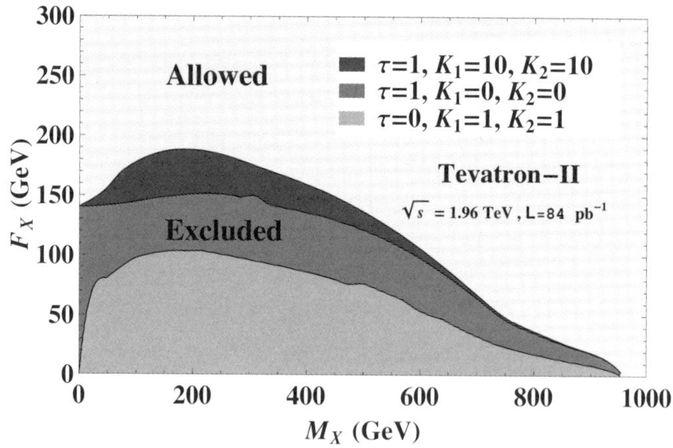

Fig. 3. Tevatron-II excluded regions of brane vector parameter space for fixed values of τ, K_1 and K_2.

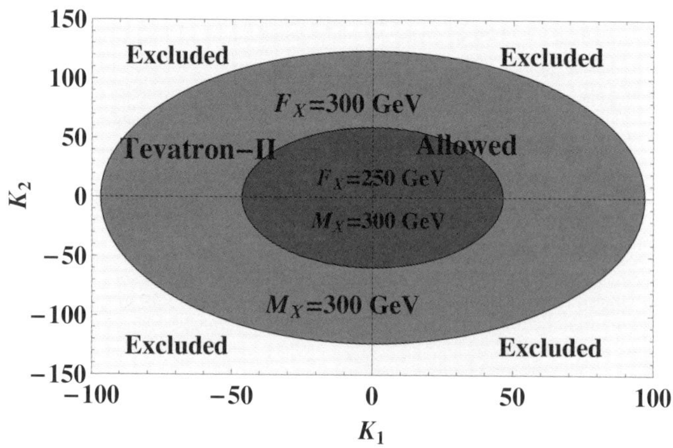

Fig. 4. Tevatron-II excluded regions of brane vector parameter space for fixed values of F_X, M_X and $\tau = 1$.

The brane vectors couple to the Standard Model in pairs and thus are stable. They are candidates for dark matter. It is found that they elude direct detection since the cross section for scattering from nuclei in the non-relativistic limit goes as at least as the second power of $v/c \simeq 0.001$ and is suppressed. The bounds for F_X and M_X from the brane vector relic abundance however yield further excluded regions of parameter space.[7]

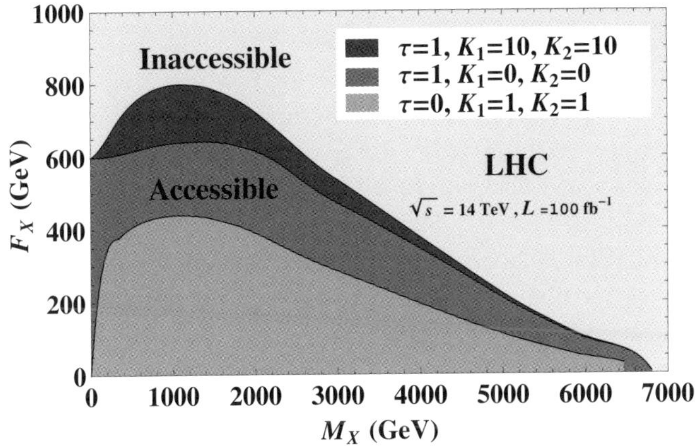

Fig. 5. LHC accessible regions of brane vector parameter space for fixed values of τ, K_1 and K_2.

Also, the brane vectors have direct coupling to the Higgs sector of the Standard Model given by the contribution to the action

$$\Gamma_{\text{HiggsXX}} = \int d^4x \frac{-1}{4F_X^2} \left[h_1 F^i_{\mu\nu} F^{i\mu\nu} + h_2 F^i_{\mu\nu} \tilde{F}^{i\mu\nu} + h_3 K^i_{\mu\nu} K^{i\mu\nu} \right] \Phi^\dagger \Phi, \tag{16}$$

with Φ the Higgs doublet. Shifting the Higgs field by its vacuum expectation value $v/\sqrt{2}$, $\Phi^\dagger\Phi \to vH$, leads to its direct decay into 2 brane vectors. This invisible decay rate can be comparable to that of the Standard Model Higgs decay rate for the allowed region of parameter space and Higgs masses approximately 120-180 GeV.[8]

Finally, this is an effective theory only approximating the brane dynamics up to a cut-off scale. Above the cut-off, the ultra violet completion of the theory is necessary to accurately reflect the dynamics, whatever that might be. This scale can be estimated by the unitarity bound for collider production of the brane vectors. The values of the effective brane tension cannot be too low or the cross section to produce 4 X particles will be the same as 2 X particles. A crude estimate of 4 X production being less than 2 X production yields a relation $(7-10)F_X = M_X$. This indicates a region of applicability of the effective theory which this line bounds below in F_X-M_X parameter space plot.[6]

For flexible brane world models a massive vector field is present in the low energy spectrum. It couples to the Standard Model fields via their energy-momentum tensor and through the extrinsic curvature of the brane.

This provides a means to delimit the brane vector parameter space by collider missing energy production processes. The focus has been the production of a single photon with missing energy. Likewise a single Z or jet plus missing energy also offers additional parameter constraints. It is possible for the Higgs particle to invisibly decay into a pair of brane vectors at rates comparable to that of the Standard Model decay modes. The brane vectors are stable and hence are candidates for dark matter. Although they elude direct detection bounds, their cosmological relic abundance places further restrictions on the collider delineated brane vector parameter space.

This work was supported in part by the U.S. Department of Energy under grant DE-FG02-91ER40681 (Task B). I would like to thank my collaborators, Sherwin Love, Muneto Nitta, Tonnis ter Veldhuis and Chi Xiong, for their valuable insight and enjoyable discussions during our work together on this research.

References

1. V. A. Rubakov and M. E. Shaposhnikov, Phys. Lett. B **125**, 136 (1983); K. Akama, Lect. Notes Phys. **176**, 267 (1982) [arXiv:hep-th/0001113]; M. Visser, Phys. Lett. B **159**, 22 (1985) [arXiv:hep-th/9910093]; G. R. Dvali and M. A. Shifman, Phys. Lett. B **396**, 64 (1997) [Erratum-ibid. B **407**, 452 (1997)] [arXiv:hep-th/9612128]; L. Randall and R. Sundrum, Phys. Rev. Lett. **83**, 4690 (1999) [arXiv:hep-th/9906064]; A. Karch and L. Randall, JHEP **0105**, 008 (2001) [arXiv:hep-th/0011156].
2. S. R. Coleman, J. Wess and B. Zumino, Phys. Rev. **177**, 2239 (1969); C. G. Callan, S. R. Coleman, J. Wess and B. Zumino, Phys. Rev. **177**, 2247 (1969); D. V. Volkov, Sov. J. Particles and Nuclei **4**, 3 (1973); V. I. Ogievetsky, Proceedings of the X-th Winter School of Theoretical Physics in Karpacz, vol. 1, p. 227 (Wroclaw, 1974).
3. T. E. Clark, S. T. Love, M. Nitta, T. ter Veldhuis and C. Xiong, Phys. Rev. D **75**, 065028 (2007) [arXiv:hep-th/0612147]; T. E. Clark, S. T. Love, M. Nitta and T. ter Veldhuis, Phys. Rev. D **72**, 085014 (2005) [arXiv:hep-th/0506094].
4. T. Kugo and K. Yoshioka, Nucl. Phys. B **594**, 301 (2001) [arXiv:hep-ph/9912496]; M. Bando, T. Kugo, T. Noguchi and K. Yoshioka, Phys. Rev. Lett. **83**, 3601 (1999) [arXiv:hep-ph/9906549].
5. T. E. Clark, S. T. Love, M. Nitta, T. ter Veldhuis and C. Xiong, "Brane Vector Phenomenology," arXiv:0709.4023 [hep-th].
6. T. E. Clark, S. T. Love, M. Nitta, T. ter Veldhuis and C. Xiong, "Brane Oscillations At The Tevatron and LHC", *Pheno 08: LHC Turn On*, Phenomenology Symposium April 2008, Madison, WI, and in preparation; "Brane Oscillations In Collider Physics", *Pheno 07: Prelude to*

the LHC, Phenomenology Symposium May, 2007, Madison, WI; "Lepton Colliders and Brane Vector Phenomenology", in preparation. See also http://www.pheno.info/symposia.
7. T. E. Clark, S. T. Love, M. Nitta, T. ter Veldhuis and C. Xiong, "Brane Vector Dark Matter", *Pheno 08: LHC Turn On*, Phenomenology Symposium April 2008, Madison, WI, and in preparation.
8. T. E. Clark, Boyang Liu, S. T. Love, T. ter Veldhuis and C. Xiong, "Higgs Decays and Brane Gravi-vectors", *Pheno 08: LHC Turn On*, Phenomenology Symposium April 2008, Madison, WI, and in preparation.
9. E. A. Ivanov and V. I. Ogievetsky, Teor. Mat. Fiz. **25**, 164 (1975).
10. A. M. Polyakov, Nucl. Phys. B **268**, 406 (1986).
11. Y. Nambu, Phys. Rev. D **10**, 4262 (1974); T. Goto, Prog. Theor. Phys. **46**, 1560 (1971).
12. T. E. Clark, S. T. Love, M. Nitta, T. ter Veldhuis and C. Xiong, Phys. Rev. D **76**, 105014 (2007) [arXiv:hep-th/0703179]; T. E. Clark, S. T. Love, M. Nitta and T. ter Veldhuis, J. Math. Phys. **46**, 102304 (2005) [arXiv:hep-th/0501241]; T. E. Clark, M. Nitta and T. ter Veldhuis, Phys. Rev. D **67**, 085026 (2003) [arXiv:hep-th/0208184].
13. P. Creminelli and A. Strumia, Nucl. Phys. B **596**, 125 (2001) [arXiv:hep-ph/0007267].
14. J. A. R. Cembranos, A. Dobado and A. L. Maroto, Phys. Rev. D **70**, 096001 (2004) [arXiv:hep-ph/0405286]; Phys. Rev. D **73**, 035008 (2006) [arXiv:hep-ph/0510399]; Phys. Rev. D **73**, 057303 (2006) [arXiv:hep-ph/0507066]; Phys. Rev. D **68**, 103505 (2003) [arXiv:hep-ph/0307062]; Phys. Rev. Lett. **90**, 241301 (2003) [arXiv:hep-ph/0302041]; J. Alcaraz, J. A. R. Cembranos, A. Dobado and A. L. Maroto, Phys. Rev. D **67**, 075010 (2003) [arXiv:hep-ph/0212269]; A. Dobado and A. L. Maroto, Nucl. Phys. B **592**, 203 (2001) [arXiv:hep-ph/0007100].
15. T. E. Clark, S. T. Love, M. Nitta, T. ter Veldhuis and C. Xiong, "Brane Vector Dynamics from Embedding Geometry", in preparation; "Embedding geometry and decomposition of gravity", T.E. Clark, S.T. Love, T. ter Veldhuis and C. Xiong, in *Proceedings of The Fourth Meeting on CPT and Lorentz Symmetry*, Bloomington, 2007, (World Scientific, Singapore 2008) pp. 260-264.
16. P. Achard *et al.* [L3 Collaboration], Phys. Lett. B **597**, 145 (2004) [arXiv:hep-ex/0407017]; S. Mele, "Search for Branons at LEP", Int. Europhys. Conf. on High Energy Phys., PoS(HEP2005)153.
17. W. K. Tung, H. L. Lai, A. Belyaev, J. Pumplin, D. Stump and C. P. Yuan, JHEP **0702**, 053 (2007) [arXiv:hep-ph/0611254]; see also http://hep.pa.msu.edu/cteq/public/cteq6.html, http://www.phys.psu.edu / cteq/ and http://durpdg.dur.ac.uk/hepdata/pdf3.html.
18. D. Acosta [CDF Collaboration], Phys. Rev. Lett. **92**, 121802 (2004) [arXiv:hep-ex/0309051] and Phys. Rev. Lett. **89**, 281801 (2002); P. Onyisi, "A Search for New Physics in the Exclusive Photon and Missing E_T Channel at CDF", Univ. of Chicago/CDF, APS April 2003; V. M. Abazov *et al.* [D0 Collaboration], arXiv:0803.2137 [hep-ex].

KNOTS AS POSSIBLE EXCITATIONS OF THE QUANTUM YANG-MILLS FIELDS

LUDWIG D. FADDEEV*

St. Petersburg Department of the Steklov Mathematical Institute
St. Petersburg, Russia
**E-mail: faddeev@euclid.pdmi.ras.ru*

Keywords: Yang-Mills theory; Knot theory

1. Dedication

It is a great pleasure to contribute to the volume, dedicated to 80 years celebration of Professor Zimmermann, my long time senior colleague in the Quantum Field Theory. I hope, that my text combines traditional methods of this theory with some more modern ideas.

2. Introduction

Quantum Yang-Mills theory[1] is most probably the only viable relativistic field theory in 4-dimensional space-time. The special property, leading to this conviction, is dimensional transmutation[2] and related property of asymtotic freedom.[3] However the problem of description of corresponding particle-like excitations is still not solved. The question, posed by W. Pauli in 1954 during talk of C. N. Yang at Oppenheimer seminar at IAS,[4] waits for an answer for more than 50 years. In this talk I shall present a hypothetical scenario for this picture: particles of Yang-Mills field are knot-like solitons.

The idea is based on another popular hypothese, according to which the confinement in QCD is effectuated by gluonic strings, connecting quarks. Thus a natural question is what happenes to these strings in the absence of quarks, i.e. in the pure Yang-Mills theory. The strings should not disappear, they rather become closed, producing rings, links or knots. This idea was leading in my recent activity in collaboration with Antti Niemi.

Our approach is based on a soliton model, which I proposed in the mid-70ties in the wake of interest to the soliton mechanism for particle-like excitations. My proposal was mentioned in several talks, partly refered to in.[5]

The model is a kind of nonlinear σ-model with nonlinear field $n(x)$ taking values in the two-dimensional sphere $\mathbb{S}^2$. It does not allow complete separation of variables, so practical research was to wait until mid-90ties when computers strong enough became available. It was Antti Niemi, who was first to sacrifice himself for complicated numerical work with the great help of supercomputer center at Helsinki. The first result, published in,[5] attracted attention of two groups.[6,7] Their work revealed rich structure of knot-like solitons, confirming my expectations. Thus a candidate for dynamical model with knot-like excitations was found. Next step was to find a place for this field among the dynamical variables of the Yang-Mills field theory.

We developed consequentively two approaches for this. The first one was based on the proposal of Y. M. Cho[8] to construct kind of the magnetic monopole connection, described by means of the n-field.[9] This approach is still discussed by several groups.[10,11] In fact Cho connection was found before in.[12] However, now we do not consider this approach as promising anymore and in the beginnning of new century developed another one. The short announcement[13] was developed in a detailed paper.[14] In this talk I shall briefly describe our way to this proposal and give its exposition. I shall begin with the description of the σ-model, then propose its application in the condenced matter theory and finally explain our approach to the Yang-Mills theory.

3. Nonlinear σ-model

The field variable is n-field — a unit vector

$$\vec{n} = (n_1, n_2, n_3), \quad \sum n_i^2 = 1.$$

In other words the target is a sphere $\mathbb{S}^2$. For static configurations the space variables ran through $\mathbb{R}^3$. Boundary condition

$$n|_\infty = (0, 0, 1)$$

compactifies $\mathbb{R}^3$ to $\mathbb{S}^3$, so n-field realizes the map

$$n : \mathbb{S}^3 \to \mathbb{S}^2.$$

Such maps are classified by means of the topological charge, called Hopf invariant, which is more exotic in comparison with more usual degree of map, used when space and target have the same dimension.

To describe this topological charge consider the preimage of the volume form on $\mathbb{S}^2$ — 2 form on $\mathbb{R}^3$ (or $\mathbb{S}^3$)

$$H = H_{ik} dx^i \wedge dx^k$$

with

$$H_{ik} = (\partial_i \vec{n} \times \partial_i \vec{n}, \vec{n}) = \epsilon_{abc} \partial_i n^a \partial_k n^b n^c,$$

which is exact

$$H = dC.$$

Then Chern-Simons integral

$$Q = \frac{1}{4\pi} \int_{\mathbb{R}^3} H \wedge C$$

acquires only integer values and is called Hopf invariant.

The formulas above have natural interpretation in terms of magnetic field. Indeed, the Poincare dual of H_{ik}

$$B_i = \frac{1}{2} \epsilon_{ikj} H_{kj}$$

is divergenceless

$$\partial_i B_i = 0$$

and can be taken as a description of magnetic field. The preimage of a point on $\mathbb{S}^2$ is a closed contour, describing a line of force of this field. The Hopf invariant is an intersection number of any two such lines.

It is instructive to mention, that n-filed gives a way to describe the magnetic field alternative to one based on the the vector potential. In particular the configuration

$$\vec{n} = \frac{\vec{x}}{|x|}$$

describes the magnetic monopole without annoying Dirac string.

There are two natural functional, which can be used to introduce the energy. The first is the traditional σ-model hamiltonian

$$E_1 = \int_{\mathbb{R}^3} (\partial n)^2 d^3 x.$$

The second is the Maxwell energy of magnetic field

$$E_2 = \int_{\mathbb{R}^3} (H_{ik})^2 d^3x.$$

Functional E_1 is quadratic in the derivatives of n-field and E_2 is quartic in them. Thus they have opposite reaction to scaling $x \to \lambda x$

$$E_1 \to \lambda E_1, \quad E_2 = \frac{1}{\lambda} E_2,$$

which is reflected in their different dimensions

$$[E_1] = [L], \quad [E_2] = [L]^{-1}.$$

We take for the energy their linear combination

$$E = aE_1 + bE_2,$$

where

$$[a] = [L]^{-2}$$

and b is dimensionless. Derric theorem – the well known obstruction for the existence of localized finite energy solutions (solitons) – does not apply here.

The estimate

$$E \geq c|Q|^{3/4},$$

found in,[15] supports the belief that such solutions do exist. Unfortunately the relevant mathematical theorem is not proved until now, so we are to refer to numerical evidence.[6,7] The picture of solutions looks as follows. The lowest energy $Q = 1$ soliton is axial symmetric; it is concentrated along the circle $n_3 = -1$; the magnetic surfaces (preimages of lines $n_3 = \text{const}$) are toroidal, wrapped once by by magnetic lines of force. For $Q = 4$ minimal solution is a link and for $Q = 7$ it is trefoil. Beautiful computer movies, illustrating the calculations based on the descent method, can be found in.[16]

There is a superficial analogy of the σ-model with the Skyrme model[17] for the principal chiral field $g(x)$ with values in the manifold of compact Lie group G. Skyrme lagrangian is expressed via the Maurer-Cartan current

$$L_\mu = \partial_\mu g g^{-1}$$

as follows

$$\mathcal{L} = a \,\mathrm{tr}\, L_\mu^2 + b \,\mathrm{tr}[L_\mu, L_\nu]^2,$$

which also contains terms quadratic and quartic in derivatives of g. Corresponding topological charge

$$Q = \int \mathrm{tr}[L_i, L_k] L_j \epsilon_{ikj} d^3x$$

coinsides with the degree of map for $G = SU(2)$. There is an estimate for static Hamiltonian

$$E \geq c|Q|.$$

The minimal excitation for $Q = 1$ is spherically symmetric and concentrated around a point.

So there are two important differences between two models. First, the excitations of Skyrme model are point like, whereas those for nonlinear σ-model are string-like. Second the term E_2 has natural interpretation as Maxwell energy whereas the quartic term in the Skyrme model is rather artificial.

This concludes the description of the nonlinear σ-model and I must turn to its applications. Before the main one to Yang-Mills field, I shall consider more simple example, developed together with Niemi and Babaev.[18]

4. Two component Landau-Ginzburg-Gross-Pitaevsky equation

The equation from the title appears in the theory of superconductivity (LG) and Bose gas (GP). The main degree of freedom is a complex valued function $\psi(x)$ — gap in the superconductivity or density in Bose-gas. Magnetic field is described by vector potential $A_k(x)$. There is a huge literature dedicated to the LGGP equation. Our contribution consists in using two components ψ

$$\psi = (\psi_1, \psi_2),$$

corresponding to a mixture of two materials.

The energy in the appropriate units is written as

$$E = \sum_{\alpha=1}^{2} |\nabla_i \psi_\alpha|^2 + F_{ik}^2 + v(\psi),$$

where

$$\nabla_i \psi = \partial_i \psi + i A_i \psi$$

and

$$F_{ik} = \partial_i A_k - \partial_k A_i.$$

The functional E is invariant with respect to the abelian gauge transformation

$$A_i \to A_i + \partial_i \lambda, \quad \psi_\alpha \to e^{-i\lambda}\psi_\alpha$$

with an arbitrary real function λ. In the case of one component ψ the change of variables

$$\psi = \rho e^{i\theta}, \quad A_k = B_k + \frac{1}{\rho^2} J_k,$$

where

$$J_k = \frac{1}{2i}\left(\overline{\psi}\partial_k \psi - \overline{\partial_k \psi}\psi\right),$$

transforms E to the gauge invariant form

$$E = (\partial\rho)^2 + \rho^2 B^2 + (\partial_i B - \partial_k B)^2 + v(\rho),$$

eliminating phase θ and leaving gauge invariant density ρ and supercurrent B. The potential $v(\rho)$ is supposed to produce the nonzero mean value for ρ

$$<\rho>= \Lambda,$$

vector field B becomes massive (Meissner effect with finite penetration length).

In the case of two components $\psi_\alpha, \alpha = 1, 2$ the analogous change of variables, proposed in,[18] looks as follows

$$\rho^2 = |\psi_1|^2 + |\psi_2|^2,$$

$$\vec{n} = \frac{1}{\rho^2}(\bar{\psi_1}, \bar{\psi_2})\vec{\tau}\begin{pmatrix}\psi_1 \\ \psi_2\end{pmatrix}$$

$$A_k = B_k + \frac{1}{\rho^2} J_k$$

$$J_k = \frac{1}{2i}\sum_\alpha \left(\bar{\psi}_\alpha \partial_k \psi_\alpha - \partial_k \bar{\psi}_\alpha \psi_\alpha\right).$$

Here $\vec{\tau} = (\tau_1, \tau_2, \tau_3)$ is set of Pauli matrices

$$\tau_1 = \begin{pmatrix}0 & 1 \\ 1 & 0\end{pmatrix}, \quad \tau_2 = \begin{pmatrix}0 & -i \\ i & 0\end{pmatrix}, \quad \tau_3 = \begin{pmatrix}1 & 0 \\ 0 & -1\end{pmatrix}.$$

Variables ρ, B and n are gauge invariant. Thus the difference with the case of one component is appearence of the n-field.

The energy in new variables looks as follows

$$E = (\partial\rho)^2 + \rho^2(\partial n)^2 + \left(\partial_i B_k - \partial_k B_i + H_{ik}\right)^2 + \rho^2 B^2 + v(\rho)$$

and contains both ingredients of the nonlinear σ-model from section 1. If due to the Meissner effect massive vector field B vanishes in the bulk, only n-field remains there and should produce knot-like excitations. This is our main prediction and we wait for the relevant experimental work.

I want to stress the difference of our excitations with Abrikosov vortices. Our closed strings have finite energy in 3-dimensional bulk, whereas Abrikosov vortices are two-dimensional. Moreover, the corresponding topological charges are distinct — Hopf invariant in our case and degree of map $\mathbb{S}^1 \to \mathbb{S}^1$ in the case of Abrikosov vortices.

Now it is time to turn to the main subject — Yang-Mills field.

5. $SU(2)$ Yang-Mills theory

The field variables are 3 vector fields $A^a_\mu, a = 1, 2, 3$, describing connection in the fiber bundle $M_4 \times SU(2)$, where M_4 is a space-time, which for definiteness we shall take as euclidean $\mathbb{R}^4$. Let τ^a be Pauli matrices and

$$A_\mu = A^a_\mu \tau^a.$$

The gauge tranformation is given by

$$A_\mu \to g A_\mu g^{-1} + \partial_\mu g g^{-1}$$

with arbitrary 2×2 unitary matrix g. The curvature (field strength) $F_{\mu\nu}$

$$F_{\mu\nu} = \partial_\mu A_\nu - \partial_\nu A_\mu + [A_\mu, A_\nu]$$

transforms homogeneously

$$F_{\mu\nu} \to g F_{\mu\nu} g^{-1}$$

and Largangian

$$\mathcal{L}_{\mathrm{YM}} = \frac{1}{4} \operatorname{tr}(F_{\mu\nu})^2$$

is gauge invariant.

The maximal abelian partial gauge fixing (MAG), which we shall use, put restriction on the offdiagonal components A^1_μ and A^2_μ. We shall use the complex combination

$$B_\mu = A^1_\mu + i A^2_\mu$$

and MAG condition looks as follows

$$\nabla_\mu B_\mu = 0,$$

where

$$\nabla_\mu = \partial_\mu + iA_\mu, \quad A_\mu = A^3_\mu.$$

The fact, that we use a distinguished (diagonal) direction in the charge space is not essential, see[14] for details.

The remaining gauge freedom is the abelian one

$$B_\mu \to e^{-i\lambda} B_\mu, \quad A_\mu \to A_\mu + \partial_\mu \lambda.$$

MAG condition can be realized by adding the quadratic form $\frac{1}{2}|\nabla_\mu B_\mu|^2$ to $\mathcal{L}_{\mathrm{YM}}$, leading to

$$\mathcal{L}_{\mathrm{MAG}} = \mathcal{L}_{\mathrm{YM}} + \frac{1}{2}|\nabla_\mu B_\mu|^2$$
$$= \frac{1}{2}|\nabla_\mu B_\nu|^2 + \frac{1}{4}(F_{\mu\nu} + H_{\mu\nu})^2 + \frac{1}{2}F_{\mu\nu}H_{\mu\nu},$$

where

$$F_{\mu\nu} = \partial_\mu A_\nu - \partial_\nu A_\mu, \quad H_{\mu\nu} = \frac{1}{2i}(\bar{B}_\mu B_\nu - \bar{B}_\nu B_\mu).$$

The last term appears after the integration by parts, used to eliminate the unwanted term $\bar{\nabla}_\mu \bar{B}_\nu \nabla_\nu B_\mu$.

Now I come to the main trick. Observe, that two vector fields A^1_μ and A^2_μ define 2-plane in M_4. Let us parametrize this 2-plane by the orthogonal zweibein e_μ

$$e_\mu = e^1_\mu + ie^2_\mu$$
$$\bar{e}_\mu e_\mu = 1, \quad e^2_\mu = \bar{e}^2_\mu = 0$$

and express B_μ as

$$B_\mu = \psi_1 e_\mu + \psi_2 \bar{e}_\mu,$$

introducing two complex coefficients ψ_1 and ψ_2. Altogether the set e_μ, ψ_1, ψ_2 contains 9 real functions and B_μ has only 8 real components. The discrepancy is resolved by comment, that expression for B_μ is invariant with respect to the abelian gauge transformation

$$e_\mu \to e^{i\omega} e_\mu, \quad \psi_1 \to e^{-i\omega}\psi_1, \quad \psi_2 \to e^{i\omega}\psi_2.$$

Corresponding $U(1)$ connection is given by

$$\Gamma = \frac{1}{i}(\bar{e}_\nu \partial_\mu e_\nu), \quad \Gamma_\mu \to \Gamma_\mu + \partial_\mu \omega.$$

Now having ψ_1, ψ_2 we can repeat trick from section 2, introducing n-field. However in our case we can do more. Indeed, the combination $H_{\mu\nu}$, entering $\mathcal{L}_{\mathrm{MAG}}$, can be written as

$$H_{\mu\nu} = \rho^2 n_3^2 g_{\mu\nu}$$

with

$$\rho^2 n_3^2 = |\psi_1|^2 - |\psi_2|^2, \quad g_{\mu\nu} = \frac{1}{2i}(\bar{e}_\mu e_\nu - \bar{e}_\nu e_\mu).$$

Putting

$$p_i = g_{0i}, \quad q_i = \frac{1}{2}\epsilon_{ijk} g_{jk}$$

we get two vectors p_i, q_i satisfying conditions

$$p^2 + q^2 = 1, \quad (p,q) = 0,$$

thus defining two spheres $\mathbb{S}^2$. Indeed, what we get here is a particular parametrization of the Grassmanian $G(4,2)$. In static case p_i disappears and we are left with one unit 3-vector q, which evidently could be used to introduce the magnetic monopoles.

Now we can put the new variables into $\mathcal{L}_{\mathrm{MAG}}$. All details are to be found in.[14] Here I shall write explicitely the static energy

$$\begin{aligned} E &= (\partial_i \rho)^2 + \rho^2\big((\nabla_k n)^2 + (\partial_k q)^2\big) + \rho^2 C_k^2 \\ &+ \frac{1}{4}\big((\partial_i n \times \partial_k n, n) + (\partial_i q \times \partial_k q, q) + 2H_{ik} + \partial_i C_k - \partial_k C_i\big)^2 - \frac{3}{4}\rho^4 n_3^2, \end{aligned}$$

where C is supercurrent

$$C_k = A_k + \frac{1}{2\rho^2}\big(\bar{\psi}_1(\partial_k + iA_k + i\Gamma_k)\psi_1 + \bar{\psi}_2(\partial_k + iA_k - i\Gamma_k)\psi_2 - \text{c.c.}\big)$$

and

$$\nabla_k n^a = \partial_k n^a + \epsilon^{ab3}\Gamma_k n^b.$$

We see, that the structure of nonlinear σ-model appears twice — via fields n and q. We can interprete it as a new manifestation of electromagnetic duality in the nonabelian Yang-Mills theory.

The expression for E can be taken as a point of departure for speculations on the knot-like excitations for the $SU(2)$ Yang-Mills field. The corresponding transformation for $SU(3)$ case, done in,[19] is more complicated due to difference of rank and number of roots.

I want to stress, that by no means I propose to use the new variables to make a change of variables in the functional integral. Rather they should be put into the renormalized effective action, which should be found in the background field formalism. The variant of this method, where the background field is not classical, but is a solution of the quantum modified equation of motion is given in.[20] In the course of renormalization this effective action should experience the dimensional transmutation. We still do not know, how it happens, so we can only use speculations. The main hope, that in this way the mean value of ρ^2 will appear. At the same time the classical lagrangian should be main part of the effective action, as it represents the only possible local gauge invariant functional of dimension -4. The condensate of $< \rho^2 >$ of dimension -2 is the subject of many papers in the last years (see e.g.[21,22]). Thus all this makes the picture of string-like excitations for the Yang-Mills field more feasible. However the real work only begins here. I hope, that this subject will take fancy of some more young researchers.

References

1. C. N. Yang and R. Mills, "Conservation of Isotopic Spin and Isotopic Gauge Invariance," *Phys. Rev.* **96**, 191–195, (1954).
2. S. Coleman, "Secret Symmetries: An Introduction to Spontaneous Symmetry Breakdown and Gauge Fields," Lecture given at 1973 Intern. Summer School in Phys. Ettore Majorana. Erice (Sicily), 1973, *Erice Subnucl. Phys.*, 1973.
3. D. J. Gross and F. Wilczek, "Ultraviolet Behavior of non-abelian Gauge Theories," Phys. Rev. Lett. **30**, 1343 (1973); H. D. Politzer, "Reliable Perturbative Results for Strong Interactions?," Phys. Rev. Lett. **30**, 1346 (1973).
4. See commentary by C. N. Yang in C. N. Yang, *Selected papers 1945-1980*, Freeman and Company, (1983) 19-21.
5. L. D. Faddeev and A. J. Niemi, "Knots and particles," Nature **387**, 58 (1997) [arXiv:hep-th/9610193]
6. J. Hietarinta, P. Salo, "Faddeev-Hopf Knots: Dynamics of Linked Unknots," Phys. Lett., 1999, **B451**, 60.
7. R. Battye, P. M. Sutcliffe, "Knots as Stable Soliton Solutions in a Three-Dimensional Classical Field Theory," Phys. Rev. Lett., 1998, **81**, 4798.
8. Y. M. Cho, "A Restricted Gauge Theory," Phys. Rev. D **21**, 1080 (1980).
9. L. D. Faddeev and A. J. Niemi, "Partially dual variables in SU(2) Yang-Mills theory," Phys. Rev. Lett. **82**, 1624 (1999) [arXiv:hep-th/9807069].
10. K. I. Kondo, T. Murakami and T. Shinohara, "Yang-Mills theory constructed from Cho-Faddeev-Niemi decomposition," Prog. Theor. Phys. **115**, 201 (2006) [arXiv:hep-th/0504107].
11. Y. M. Cho, "Knot topology of QCD vacuum," Phys. Lett. B **644**, 208 (2007) [arXiv:hep-th/0409246].
12. Y. S. Duan and M. L. Ge, Sinica Sci. **11** (1979) 1072.
13. L. D. Faddeev and A. J. Niemi, "Decomposing the Yang-Mills field," Phys. Lett. B **464**, 90 (1999) [arXiv:hep-th/9907180].

14. L. D. Faddeev and A. J. Niemi, "Spin-charge separation, conformal covariance and the SU(2) Yang-Mills theory," Nucl. Phys. B **776**, 38 (2007) [arXiv:hep-th/0608111].
15. A. F. Vakulenko, L. V. Kapitansky, *Dokl. Akad. Nauk USSR*, **248**, (1979), 840–842.
16. http://users.utu.fi/hietarin/knots/index.html
17. T. H. R. Skyrme, "A Nonlinear field theory," Proc. Roy. Soc. Lond. A **260**, 127 (1961).
18. E. Babaev, L. D. Faddeev and A. J. Niemi, "Hidden symmetry and duality in a charged two-condensate Bose system," Phys. Rev. B **65**, 100512 (2002) [arXiv:cond-mat/0106152]
19. T. A. Bolokhov and L. D. Faddeev, "Infrared variables for the SU(3) Yang-Mills field," Theor. Math. Phys. **139**, 679 (2004) [Teor. Mat. Fiz. **139**, 276 (2004)].
20. L. D. Faddeev, "Notes on divergences and dimensional transmutation in Yang-Mills theory," Theor. Math. Phys. **148**, 986 (2006) [Teor. Mat. Fiz. **148**, 133 (2006)].
21. F. V. Gubarev, L. Stodolsky and V. I. Zakharov, Phys. Rev. Lett. **86**, 2220 (2001); L Stodolsky, P. van Baal and V.I. Zakharov, Phys. Lett. **B552**, 214 (2003).
22. H. Verschelde, K. Knecht, K. Van Acoleyen and M. Vanderkelen, "The non-perturbative groundstate of QCD and the local composite operator A(mu)**2," Phys. Lett. B **516**, 307 (2001) [arXiv:hep-th/0105018].
23. K. I. Kondo, "Vacuum condensate of mass dimension 2 as the origin of mass gap and quark confinement," Phys. Lett., 2001, **B514**, 335.

FEYNMAN GRAPHS AND RENORMALIZATION IN QUANTUM DIFFUSION*

LÁSZLÓ ERDŐS*

Mathematisches Institut, Universität München
Theresienstr. 39, D-80333 München, Germany
**E-mail: lerdos@mathematik.uni-muenchen.de*

MANFRED SALMHOFER*

Institut für Theoretische Physik, Universität Leipzig, Postfach 100920, 04009 Leipzig, Germany and Max–Planck–Institut für Mathematik, Inselstr. 22, D-04103 Leipzig, Germany
**E-mail: salmhofer@itp.uni-leipzig.de*

HORNG–TZER YAU*

Mathematics Department, Harvard University, Cambridge, MA 02138, USA
**E-mail: htyau@math.harvard.edu*

We review our proof that in a scaling limit, the time evolution of a quantum particle in a static random environment leads to a diffusion equation. In particular, we discuss the role of Feynman graph expansions and of renormalization.

Keywords: Brownian motion; Anderson model; Quantum diffusion; Feynman graphs; Renormalization
AMS 2000 Subject Classification: 60J65, 81T18, 82C10, 82C44

1. Introduction

The emergence of irreversibility from reversible dynamics in large systems has been one of the fundamental questions in science since the days of Maxwell and Boltzmann. The famous debate about the statistical character of the second law of thermodynamics and the related controversy about Boltzmann's *Stoßzahlansatz* in the derivation of his transport equation has been very fruitful for physics and mathematics. After Lanford's rigorous jus-

*Talk given by M. Salmhofer at the conference in honour of Wolfhart Zimmermann's 80th birthday, Ringberg Castle, February 3–6, 2008

tification of the Boltzmann equation for a classical many–particle system at short kinetic time scales,[1] the mathematical justification of the Boltzmann equation at longer timescales has remained a challenge up to the present time. The analogous statement for quantum systems remains open even at the short kinetic timescale.

A related important question is to understand how Brownian motion emerges as an effective law from time-reversal-invariant microscopic physical laws, as given by a Hamiltonian system or the Schrödinger equation. Kesten-Papanicolaou[2] proved that the velocity distribution of a classical particle moving in an environment consisting of random scatterers (i.e., Lorenz gas with random scatterers) converges to a Brownian motion in a weak coupling limit in dimensions $d \geq 3$. In this model the bath of light particles whose fluctuations lead to the Brownian motion of the observed particle is replaced with random static impurities. A similar result was obtained in $d = 2$ dimensions.[4] Recently,[3] the same evolution was controlled on a longer time scale and the position process was proven to converge to Brownian motion as well. Bunimovich and Sinai[5] proved the convergence of the periodic Lorenz gas with a hard core interaction to a Brownian motion. In this model the only source of randomness is the distribution of the initial condition. Finally, Dürr, Goldstein and Lebowitz[6] proved that the velocity process of a heavy particle in a light ideal gas, which is a model with a dynamical environment, converges to the Ornstein-Uhlenbeck process.

Although Brownian motion was discovered and first studied theoretically in the context of classical dynamics, it also describes the motion of a quantum particle in a random environment, on a timescale that is long compared to the standard kinetic timescale.[7–9] In the following we describe this result and the strategy of the proof in a bit more detail. Besides the motivation discussed above, the random Schrödinger operator that we study is also the standard model for transport of electrons in metals with impurities, which plays a central role in the theory of the metal–insulator transition.[10,11] The outstanding open mathematical question in this area is the proof of the extended states conjecture, stating that in dimensions $d \geq 3$, at weak disorder, the spectrum of such Hamiltonians is absolutely continuous. Despite much effort, this conjecture has up to now only been proven[12–14] on the Bethe lattice, which can be interpreted as the case $d = \infty$. In a system with a magnetic field, the existence of dynamical delocalization at certain energies near the Landau levels has been proven recently.[15]

2. The problem and the main result

We consider random Schrödinger operators, both on a lattice and in the continuum, in $d \geq 3$ dimensions. In this presentation, we focus on the case $d = 3$. The time evolution of the *Anderson Model* (AM) is generated by

$$\mathrm{i}\tfrac{\partial}{\partial t}\psi(t) = H\psi(t),\ \psi(0) = \psi_0 \ \text{ with } \ H = -\frac{1}{2}\Delta + \lambda V_\omega \ \text{ on } \ell^2(\mathbb{Z}^d) \tag{1}$$

where $-\Delta$ is the standard discrete Laplacian and the potential is given by $V(x) = \sum_{a\in\mathbb{Z}^d} V_a(x)$, with $V_a(x) = v_a \delta_{x,a}$, and v_a independent identically distributed (i.i.d.) random variables. We assume that $m_k = \mathbb{E}\left(v_a^k\right)$ satisfies

$$\forall i \leq 2d : m_i < \infty, \qquad m_1 = m_3 = m_5 = 0, \quad m_2 = 1. \tag{2}$$

The continuum analogue of this model is the *Quantum Lorentz Model* (QLM), where $H = -\frac{1}{2}\Delta + \lambda V_\omega$ on $L^2(\mathbb{R}^d)$, with Δ the standard Laplacian, $V_\omega(x) = \int_{\mathbb{R}^d} B(x-y)\mathrm{d}\mu_\omega(y)$, where B is a fixed spherically symmetric Schwarz function with $0 \in \operatorname{supp}\hat{B}$, μ_ω is a Poisson point process on $\mathbb{R}^d$ with homogeneous unit density and i.i.d. random masses:

$$\mu_\omega = \sum_{\gamma=1}^{\infty} v_\gamma(\omega)\delta_{y_\gamma(\omega)}. \tag{3}$$

$\{y_\gamma(\omega)\}$ is Poisson, independent of the weights $\{v_\gamma(\omega)\}$. Again, $m_k := \mathbb{E}_v\, v_\gamma^k$ is assumed to satisfy (2).

Suppose the initial state is localized, i.e. $\hat{\psi}_0$ is smooth. How does the solution $\psi(t) = \mathrm{e}^{-\mathrm{i}tH}\psi_0$ behave for large t ? If $\lambda = 0$, the time evolution is easily calculated in Fourier space: $\hat{\psi}(t,k) = \mathrm{e}^{-\mathrm{i}te(k)}\hat{\psi}_0(k)$, with $e(k) = k^2/2$ (QLM) or $e(k) = \sum_{i=1}^d (1 - \cos k_i)$ (AM). It is equally easy to see that the motion is ballistic, i.e.

$$\langle X^2 \rangle_t = \langle \psi(t), X^2\psi(t)\rangle \sim t^2. \tag{4}$$

If $\lambda \neq 0$, one expects either localization, $\langle X^2\rangle_t = O(1)$ for all t, or diffusive behaviour (extended states), $\langle X^2\rangle_t = O(t)$, depending on λ and $\hat{\psi}_0$. The localized behaviour corresponds to dense pure point spectrum at almost every energy; this was proven for large disorder[16,17] and away from the spectrum of the Laplacian. Extended states correspond to absolutely continuous spectrum. As mentioned, the latter has been proven[12–14] on the Cayley tree for small $\lambda > 0$. At this time there is no proof of existence of extended states in $d = 3$. For a simpler case, namely that of randomness with a decaying envelopping function, i.e. $V_\omega(x) = \omega_x h(x)$, ω_x i.i.d., h fixed, there is a proof[18,19] that for $\eta > \frac{1}{2}$ and $h(x) \sim |x|^{-\eta}$ as $|x| \to \infty$, $H = -\Delta + V_\omega$ has absolutely continuous spectrum.

Our result is formulated in terms of the *Wigner function*

$$W_\psi(x,v) = \int \mathrm{d}y\, \mathrm{e}^{\mathrm{i}vy}\overline{\psi(x+\frac{y}{2})}\,\psi(x-\frac{y}{2}) \tag{5}$$

which can be thought of as an analogue of a phase space density (but can become negative). Its marginals are $\int W_\psi(x,v)\mathrm{d}x = |\hat{\psi}(v)|^2$ and $\int W_\psi(x,v)\mathrm{d}v = |\psi(x)|^2$. Moreover,

$$\hat{W}_\psi(\xi,v) = \int \mathrm{d}x\, \mathrm{e}^{-\mathrm{i}x\xi} W_\psi(x,v) = \overline{\hat{\psi}(v-\xi/2)}\,\hat{\psi}(v+\xi/2). \tag{6}$$

On the lattice, one has to modify the definition of the Wigner transform slightly.[9]

The *kinetic scaling* is given by

$$\eta = \lambda^2, \qquad \mathcal{T} = \eta t, \qquad \mathcal{X} = \eta x, \tag{7}$$

i.e. the microscopic time and space variables both become of order λ^{-2}, so that velocities remain unscaled.

Theorem 2.1.

$$\mathbb{E}W^{\eta}_{\psi(\mathcal{T}\eta^{-1})}(\mathcal{X},\mathcal{V}) \underset{\eta\to 0}{\longrightarrow} F(\mathcal{X},\mathcal{V},\mathcal{T}), \tag{8}$$

F the solution of the linear Boltzmann equation

$$\begin{aligned}&\frac{\partial}{\partial\mathcal{T}}F(\mathcal{X},\mathcal{V},\mathcal{T}) + (\nabla e)(\mathcal{V})\cdot\nabla_{\mathcal{X}}F(\mathcal{X},\mathcal{V},\mathcal{T})\\ &= 2\pi\int \mathrm{d}\mathcal{U}\,\delta(e(\mathcal{U})-e(\mathcal{V}))\left|\hat{B}(\mathcal{U}-\mathcal{V})\right|^2 [F(\mathcal{X},\mathcal{U},\mathcal{T})-F(\mathcal{X},\mathcal{V},\mathcal{T})]\,.\end{aligned} \tag{9}$$

This theorem was first proven for the continuum for small time $\mathcal{T}$,[20] then for arbitrary time,[21] and later extended to the lattice case.[22]

The *diffusive scaling* is defined by

$$\varepsilon = \lambda^{2+\kappa/2}, \qquad X = \varepsilon x, \qquad T = \varepsilon\lambda^{\kappa/2}t = \lambda^{\kappa+2}t. \tag{10}$$

This is long compared to the kinetic timescale: the kinetic variables $\mathcal{X}$ and $\mathcal{T}$ diverge as $\lambda\to 0$ when X and T are kept fixed,

$$\mathcal{X} = \lambda^{-\kappa/2}X, \qquad \mathcal{T} = \lambda^{-\kappa}T. \tag{11}$$

A first hint at diffusion is that under this scaling $\mathcal{X}^2/\mathcal{T} = X^2/T$ is independent of λ. The result for the Anderson model is

Theorem 2.2. *Let $d = 3$, $\psi_0 \in \ell^2(\mathbb{Z}^3)$ and $\psi(t)$ be the solution to the random Schrödinger equation with initial condition ψ_0. If $\kappa > 0$ is small*

enough and $\varepsilon = \lambda^{2+\kappa/2}$, *then in the limit* $\lambda \to 0$, $\mathbb{E}W^{\varepsilon}_{\psi(\lambda^{-2-\kappa}T)}$ *converges weakly to the solution* f *of a heat equation.*

More precisely: denote $\langle F \rangle_E = \Phi(E)^{-1} \int \mathrm{d}v\ F(v)\delta(E - e(v))$, *where* $\Phi(E) = \int \mathrm{d}v\ \delta(E - e(v))$. *Let* $E \in (0,3)$ *and* $D_{ij}(E) = \frac{1}{2\pi\Phi(E)} \langle \nabla_i e\ \nabla_j e \rangle_E$, *and let* f *be the solution of the heat equation*

$$\frac{\partial}{\partial T} f(T,X,E) = \nabla_X \cdot D(E) \nabla_X\ f(T,X,E) \tag{12}$$

$$f(0,X,E) = \delta(X)\ \langle |\hat{\psi}_0|^2 \rangle_E\ . \tag{13}$$

Let $\mathcal{O}(x,v)$ *be a Schwartz function on* $\mathbb{R}^d \times \mathbb{R}^d / 2\pi\mathbb{Z}^d$. *Then*

$$\begin{aligned} \lim_{\varepsilon \to 0} \sum\nolimits_{X \in (\varepsilon\mathbb{Z}/2)^d} \int \mathrm{d}v\ \mathcal{O}(X,v)\ \mathbb{E}W^{\varepsilon}_{\psi(\lambda^{-\kappa-2}T)}(X,v) \\ = \int_{\mathbb{R}^d} \mathrm{d}X \int \mathrm{d}v\ \mathcal{O}(X,v)\ f(T,X,e(v)). \end{aligned} \tag{14}$$

The limit is uniform on $[0,T_0]$ *for any* $T_0 > 0$.

We discuss some of the ideas in the proof of this theorem in Section 3.

If $\hat{\psi}_0 \in C^1$ and λ is small enough, we have the more detailed error estimate

$$\begin{aligned} &\int \mathrm{d}v \int \mathrm{d}\xi\ \hat{\mathcal{O}}(\xi,v)\ \mathbb{E}\hat{W}^{\varepsilon}_{\psi(\lambda^{-2-\kappa}T)}(\xi,v) \\ &= \int \mathrm{d}\xi \int \Phi(E)\mathrm{d}E\ \mathrm{e}^{-\frac{T}{2}\langle \xi,\ D(E)\xi \rangle_E} \langle \hat{\mathcal{O}}(\xi,\cdot) \rangle_E\ \langle \hat{W}_{\psi_0}(\varepsilon\xi,\cdot) \rangle_E + o(1). \end{aligned} \tag{15}$$

The Boltzmann equation also gives the same diffusion equation in the long time limit, but it was itself derived from the quantum mechanical time evolution only for shorter timescales. The main difficulty in the proof is to deal with contributions that vanish for $\lambda \to 0$ under kinetic scaling, but that become important under the above–defined diffusive scaling. More technically speaking, in the Feynman expansion done to analyze the time evolution, most of these terms would even diverge under diffusive scaling if we did not renormalize the propagator.

The allowed values of κ are in an interval $[0, \kappa_0)$, where κ_0 is a universal constant. For technical reasons, κ_0 has to be chosen very small in the proof. Heuristically, i.e. ignoring many of the technical complications and assuming optimal bounds, one would expect the remainder of the renormalized Feynman graph expansion to vanish up to $\kappa_0 = 2$, and to diverge for $\kappa_0 > 2$.

The diffusive scaling leads to a diffusion on the energy shells. A diffusion mixing energy shells is expected to start at $t = \lambda^{-4}$.

An intuitive way of interpreting the expansion described below is as an expansion in the number N of collisions of the particle with the randomly (but statically) arranged obstacles represented by the potential. As compared to the previous results,[21,22] the main new feature here is that under diffusive scaling, the effective number of collisions of the particle diverges. That is, not only is it necessary to expand to an order N that diverges as $\lambda \to 0$, but also the main contribution does not come from terms with a finite number of collisions.

3. Collision histories, Feynman graphs, and ladders

We discuss some of the ideas of the proof for the example of the Anderson model, i.e. the lattice situation. For the detailed bounds of Feynman graphs, the lattice leads to a number of complications,[9] but for the presentation it is easier.

3.1. *Collision histories*

Let us start with a formal time–ordered expansion, setting $H_0 = -\frac{1}{2}\Delta$ and expanding in λV. Then $\psi(t) = \mathrm{e}^{-\mathrm{i}tH}\psi_0 = \sum_{n\geq 0}\psi^{(n)}(t)$ with

$$\psi^{(n)}(t) = (-\mathrm{i}\lambda)^n \int \mathrm{d}\mu_{n+1}(s)\mathrm{e}^{-\mathrm{i}s_n H_0} V \mathrm{e}^{-\mathrm{i}s_n H_0} \ldots V\mathrm{e}^{-\mathrm{i}s_0 H_0}\psi_0 \quad (16)$$

where $s = (s_0, \ldots, s_n)$ and

$$\mathrm{d}\mu_{n+1}(s) = \int\limits_{[0,\infty)^{n+1}} \mathrm{d}s_0 \ldots \mathrm{d}s_n\, \delta\left(t - \sum_{j=0}^{n} s_j\right). \quad (17)$$

Because $V = \sum_{a\in\mathbb{Z}^d} V_a$, it is natural to split each $\psi^{(n)}$ further, $\psi^{(n)}(t) = \sum_{\mathbf{a}_n}\psi^{(n)}_{\mathbf{a}_n}(t)$. Every sequence of obstacle labels $\mathbf{a}_n = (a_1, \ldots, a_n) \in (\mathbb{Z}^d)^n$ represents a *collision history*, and for $k \in \{1, \ldots, n-1\}$, the time variables s_k in (17) are the time differences between two subsequent collisions. The delta function in (17) enforces the constraint that these time differences, together with the propagation times s_0 before the first collision and s_n after the last one, add up to the total time t. We shall discuss convergence questions about this expansion later.

Our detailed analysis takes place in momentum space, where each V acts as a convolution operator, so that

$$\hat{\psi}_n(t,p_n) = (\text{-i})^n \int \mathrm{d}\mu_{n+1}(s) \int \prod_{j=0}^{n-1} \frac{\mathrm{d}^d p_j}{(2\pi)^d}\, \mathrm{e}^{-\mathrm{i}s_j e(p_j)} \prod_{j=1}^{n} \hat{V}(p_j\text{-}p_{j-1})\hat{\psi}_0(p_0). \quad (18)$$

Very schematically, one can represent this as follows, where each of the dashed lines represents a factor λV and each of the full lines gets a phase factor $e^{-is_j e(p_j)}$.

As the convolution formula shows, one can define a momentum flow in this graph, where the momentum change $p_j - p_{j+1}$ flows away through the dashed line. Before the disorder average, there is no translation invariance in the system, so every scattering at an obstacle changes the momentum of the particle.

3.2. *Disorder average and graphs*

Recalling (6), we have

$$\mathbb{E}\left[\hat{W}_{\psi(t)}(\xi, v)\right] = \sum_{n,n'} \sum_{\mathbf{a}_n, \mathbf{a}'_{n'}} \mathbb{E}\left[\overline{\hat{\psi}^{(n)}_{\mathbf{a}_n}(t, v - \xi/2)}\hat{\psi}^{(n')}_{\mathbf{a}'_{n'}}(t, v + \xi/2)\right]. \tag{19}$$

Note that there are now two, a priori independent, collision histories, one for ψ and one for $\bar{\psi}$. It will be part of the proof to show that, in the scaling limit we consider, the only contributions after self–energy renormalization come from the so-called ladder graphs, where the two collision histories are identical: $n = n'$ and $\mathbf{a}_n = \mathbf{a}'_n$.

Because the disorder is i.i.d., translation invariance holds for the average, which means that momentum conservation also holds for the dashed lines, which for the Anderson model simply correspond to a factor λ^2, since the second moment of the disorder was normalized to 1 in (2).

The result can be represented as a graph built of two particle lines, particle–disorder vertices, which are joined by disorder lines, and, if the randomness is non-Gaussian, disorder-disorder vertices, which correspond to the higher moments of the disorder distribution. An example is

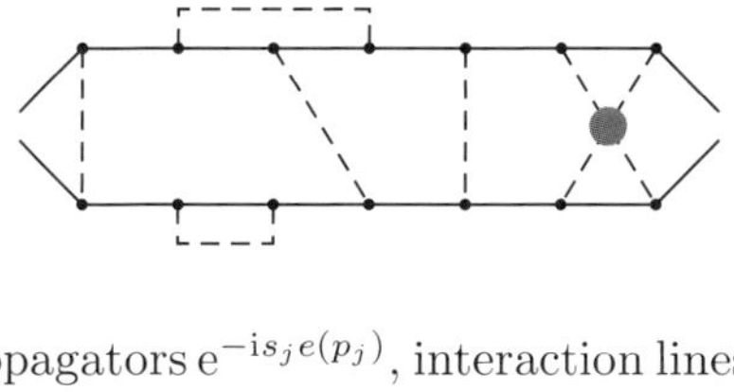

Particle lines get propagators $e^{-is_j e(p_j)}$, interaction lines give factors λ^2, and the disorder-disorder vertex of degree four corresponds to a factor $m_4 \lambda^4$.

It is clear that in the way the expansion was introduced above, one really needs the assumption that arbitrary moments, not just the first $2d$ ones, exist. The expansion employed in the true proof contains a stopping rule which avoids high moments, but we shall not discuss this here in more detail. In fact, we shall in the following assume for simplicity that the disorder is Gaussian, so that there are no vertices of higher degree for the dashed lines, and the average just corresponds to a pairing of interaction lines.

An example of a pairing is as follows

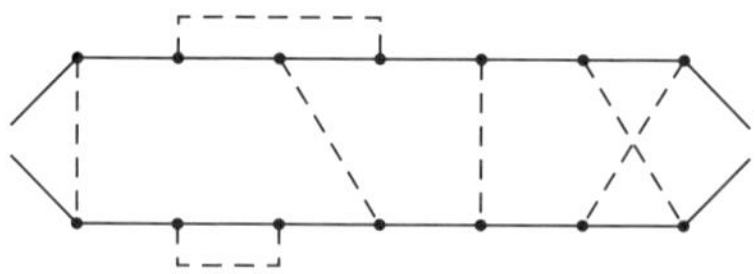

Note that here, there is a crossing of the two pairing lines in the graphical representation, but there are no vertices in which more than one interaction line enters.

A special class of pairings are the *up–down pairings*, where $n = n'$ and the pairing corresponds to a permutation $\sigma \in \mathcal{S}_n$:

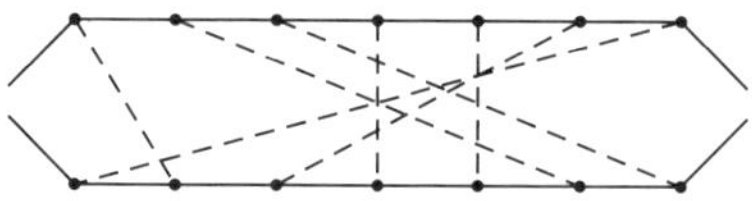

The most important term turns out to be the *ladder graph*, corresponding to $\sigma =$ id:

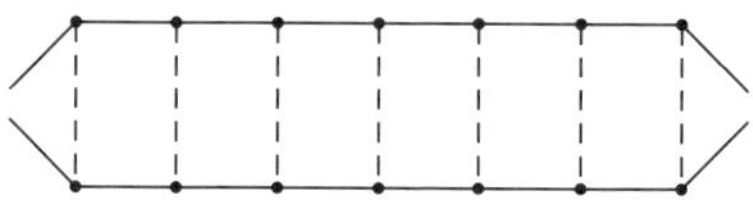

3.3. *Graph bounds*

In the following, we give a brief discussion of bounds of the contributions of individual graphs, restricting to up–down pairings. If one takes a bound in the representation (18), each phase factor is replaced by 1. This leads to a bound of order $(\lambda t)^n/n!$ (where the $n!$ comes from the time ordering implied by the delta function in (17)), which does not even allow to consider the kinetic scaling where $\lambda^2 t$ is fixed. For this reason, the following *propagator representation* is useful. Let $\eta > 0$. Then, inserting the Fourier representation of the delta function,

$$\begin{aligned}
&\int_{[0,\infty)^{n+1}} \mathrm{d}^{n+1}s \, \delta\left(t - \sum_{j=1}^{n+1} s_j\right) \prod_{j=1}^{n+1} \mathrm{e}^{-\mathrm{i}s_j e(p_j)} \\
&= \mathrm{e}^{t\eta} \int_{[0,\infty)^{n+1}} \mathrm{d}^{n+1}s \, \delta\left(t - \sum_{j=1}^{n+1} s_j\right) \prod_{j=1}^{n+1} \mathrm{e}^{-\mathrm{i}s_j (e(p_j) - \mathrm{i}\eta)} \\
&= \mathrm{e}^{t\eta} \int \frac{\mathrm{d}\alpha}{2\pi} \, \mathrm{e}^{-\mathrm{i}t\alpha} \int_{[0,\infty)^{n+1}} \mathrm{d}^{n+1}s \prod_{j=1}^{n+1} \mathrm{e}^{-\mathrm{i}s_j(\alpha - e(p_j) + \mathrm{i}\eta)} \\
&= \mathrm{i}^{-n} \mathrm{e}^{t\eta} \int \frac{\mathrm{d}\alpha}{2\pi} \, \mathrm{e}^{-\mathrm{i}\alpha t} \prod_{j=1}^{n+1} \frac{1}{\alpha - e(p_j) + \mathrm{i}\eta} \, .
\end{aligned} \tag{20}$$

It is convenient to choose $\eta = t^{-1}$.

The contribution of a permutation $\sigma \in \mathcal{S}_n$, corresponding to an up–down pairing graph Γ_σ, to $\langle \hat{\mathcal{O}}, \hat{W}^\varepsilon_\psi \rangle$ is

$$\begin{aligned}
Val(\Gamma_\sigma) = \lambda^{2n} \, \mathrm{e}^{2t\eta} & \int \frac{\mathrm{d}\alpha \, \mathrm{d}\beta}{(2\pi)^2} \, \mathrm{e}^{\mathrm{i}(\beta - \alpha)t} \\
& \int \mathrm{d}\xi \int \prod_{j=0}^{n} \frac{\mathrm{d}^d p_j}{(2\pi)^d} \int \prod_{k=0}^{n} \frac{\mathrm{d}^d q_k}{(2\pi)^d} \quad \hat{\mathcal{O}}(\xi, p_n) \hat{W}^\varepsilon_{\psi_0}(\xi, p_0) \\
& \prod_{j=0}^{n} \frac{1}{\beta - \overline{\omega(q_j - \frac{\varepsilon\xi}{2})} - \mathrm{i}\eta} \, \frac{1}{\alpha - \omega(p_j - \frac{\varepsilon\xi}{2}) - \mathrm{i}\eta} \\
& \prod_{j=1}^{n} \delta\left(p_j - p_{j-1} - (q_{\sigma(j)} - q_{\sigma(j)-1})\right) .
\end{aligned} \tag{21}$$

At the moment, $\omega(p) = e(p) \in \mathbb{R}$; later, ω will change under renormalization and become complex.

A simple Schwarz inequality separating the dependence on the p_i and that on the q_i implies that for all σ

$$|Val(\Gamma_\sigma)| \leq Val(\Gamma_{id}). \tag{22}$$

The ladder is easy to calculate at $\xi = 0$, and a ladder of length n is of order $\frac{1}{n!}(\lambda^2 t)^n = \frac{1}{n!}\mathcal{T}^n$.

A crucial observation is that the values of graphs with *crossings* get inverse powers of t, as compared to the ladder. This follows from the bound

$$\int \mathrm{d}p \, \frac{1}{|\alpha - \omega(p) + \mathrm{i}\eta|} \, \frac{1}{|\beta - \overline{\omega(\pm p + q)} - \mathrm{i}\eta|} \leq C |\log \eta|^3 \frac{\eta^{-b}}{\|q\| + \eta} \tag{23}$$

($b = 0$ for the continuum; $1/2 \leq b \leq 3/4$ on the lattice). $\|p\| = |p|$ in the continuum, $\|p\| = \min\{|p - v| : v_i \in \{0, \pm\pi\}\}$ on the lattice. Again, here $\omega(p) = e(p)$. This motivates why the ladder graph gives the dominant contribution under kinetic scaling. However, the number of graphs goes like $n!$, which cancels the $1/n!$, hence expanding to infinite order one gets majorants by geometric series, which converge only on very short kinetic timescales $\mathcal{T}$. This is the reason for the restriction to small kinetic timescales in the first proof[20] of the Boltzmann equation for the QLM.

3.4. *Expansions to finite order and remainder terms*

Major progress[21] came from the realization that one can do an expansion to finite order with an efficient remainder estimate. A natural way to generate a finite–order expansion is the *Duhamel formula*

$$\psi(t) = \mathrm{e}^{-\mathrm{i}tH} \psi_0 = \mathrm{e}^{-\mathrm{i}tH_0} \psi_0 + \int_0^t \mathrm{d}s \, \mathrm{e}^{-\mathrm{i}(t-s)H} \lambda V \mathrm{e}^{-\mathrm{i}sH_0} \psi_0. \tag{24}$$

Iteration gives

$$\psi(t) = \sum_{n=0}^{N-1} \psi^{(n)}(t) + \Psi_N(t), \tag{25}$$

where

$$\Psi_N(t) = (-\mathrm{i}) \int_0^t \mathrm{d}s \, \mathrm{e}^{-\mathrm{i}(t-s)H} \lambda V \psi^{(N-1)}(s) \tag{26}$$

and

$$\psi^{(n)}(t) = (-\mathrm{i}\lambda)^n \int \mathrm{d}\mu_{n+1}(s) \mathrm{e}^{-\mathrm{i}s_n H_0} V \ldots V \, \mathrm{e}^{-\mathrm{i}s_0 H_0} \psi_0. \tag{27}$$

An alternative way of looking at this is via its relation to the resolvent formula

$$R_z = R_z^{(0)} + R_z \lambda V R_z^{(0)} \tag{28}$$

where $R_z = (z - H)^{-1}$ and $R_z^{(0)} = (z - H_0)^{-1}$. Iteration of the resolvent equation and using the Fourier transform gives the above propagator representation directly. The Duhamel formula is obtained via

$$\mathrm{e}^{-\mathrm{i}tH} = -\mathrm{e}^{t\eta} \int \frac{\mathrm{d}\alpha}{2\pi\mathrm{i}} \, \mathrm{e}^{-\mathrm{i}\alpha t} \, R_{\alpha+\mathrm{i}\eta}. \tag{29}$$

The second crucial ingredient is that one can use the unitarity of the full time evolution to reduce all terms to ones where no H appears in the time evolution any more:

$$\|\Psi_N(t)\| \le \int_0^t \mathrm{d}s \left\| \mathrm{e}^{-\mathrm{i}(t-s)H} \lambda V \psi^{(N-1)}(s) \right\| \le \int_0^t \mathrm{d}s \left\| \lambda V \psi^{(N-1)}(s) \right\|. \quad (30)$$

Thus

$$\|\Psi_N(t)\|^2 \le t\, |\lambda|^2 \int_0^t \mathrm{d}s \left\| V \psi^{(N-1)}(s) \right\|^2. \quad (31)$$

The remaining integral over s effectively gives a factor t, which is the price to pay for this unitarity bound. To control this factor, one needs exhibit more factors t^{-1} in graphs with several independent crossings, and treat graphs with only one crossing explicitly (in the resolvent iteration, the unitarity bound would be replaced by $\|R_{\alpha+\mathrm{i}\eta}\| \le \eta^{-1}$).

By a Schwarz inequality, one can see that the Wigner transform is continuous in L^2 norm:

$$\left| \mathbb{E} \left(\langle \hat{\mathcal{O}}, \hat{W}^{\varepsilon}_{\psi_1} \rangle - \langle \hat{\mathcal{O}}, \hat{W}^{\varepsilon}_{\psi_2} \rangle \right) \right| \le C \int \mathrm{d}\xi \, \sup_v \left| \hat{\mathcal{O}}(\xi, v) \right| \sqrt{\mathbb{E}\|\psi_1\|^2 \, \mathbb{E}\|\psi_1 - \psi_2\|^2}. \quad (32)$$

Thus the unitarity bound can also be used for the Wigner transform. The proof of the Boltzmann equation[21] on an arbitrarily large kinetic timescale $\mathcal{T}$ uses an expansion up to order $N \sim \log t$. The ladder terms give the gain term in the Boltzmann equation. The lowest order self–energy correction gives the loss term in the Boltzmann equation. It corresponds to the "gate" graph

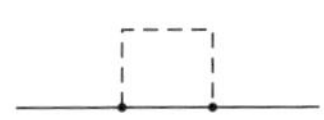

3.5. *Long time scale: Renormalization*

Because the ladder with n rungs is of order $(\lambda^2 t)^n / n!$, it diverges under diffusive scaling, and so do other graphs. To increase the time beyond λ^{-2}, we need to do a renormalization. Formally, one can think of this as a resummation of the gate diagrams, which are of self–energy type, but this geometric series converges only for small $\lambda^2 t$. A way to avoid such formal resummations is to change the way H is split into a "free" and an interaction part, i.e., expand around a different H_0. For $\varepsilon > 0$ set

$$\Theta_\varepsilon(\alpha) = \int \mathrm{d}q \frac{1}{\alpha - e(q) + \mathrm{i}\varepsilon} . \tag{33}$$

This is the value of the gate diagram at energy α in the Anderson model (in the QLM, the integrand contains an additional factor from the interaction function). The limit $\Theta(\alpha) = \lim_{\varepsilon\to 0+} \Theta_\varepsilon(\alpha)$ exists and is Hölder continuous[7] in α of order $1/2$. Let

$$\theta(p) = \Theta(e(p)). \tag{34}$$

The idea is now to put $\lambda^2\theta(p)$ as a counterterm, which by construction subtracts every insertion of a gate diagram at the point $\alpha = e(p)$ where the particle propagator is singular. Because α and η appear only as auxiliary quantities in the expansion, it was necessary to take $\varepsilon \to 0$ above and define θ in an α–independent way.

The counterterm is added and subtracted so that the Hamiltonian does not change: let $\omega(p) = e(p) + \lambda^2\theta(p)$ and decompose

$$H = \omega(P) + U, \qquad U = \lambda V - \lambda^2\theta(P) \tag{35}$$

(where P denotes the momentum operator). The function ω can be thought of as a new dispersion relation of energy as a function of momentum. However, ω also has a negative imaginary part, roughly of order λ^2. More precisely, for $d \geq 3$ there is $c > 0$ such that

$$\mathrm{Im}\ \omega(p) \leq -c\lambda^2 |\!|\!| p |\!|\!|^{d-2}. \tag{36}$$

Thus H_0 is no longer selfadjoint. However, the negative sign of $\mathrm{Im}\ \omega$ implies that the resolvent $R_{\alpha+\mathrm{i}\eta}$ is still well-defined, since the imaginary parts add up with the same sign. Correspondingly, the time evolution operator $\mathrm{e}^{-\mathrm{i}sH_0}$ is no longer unitary but it remains bounded for $s \geq 0$. Both the Duhamel and the resolvent iteration are thus well-defined. Besides the new propagator $(\alpha + \mathrm{i}\eta - \omega(p))^{-1}$, the important change is that every factor U now also contains a counterterm insertion $-\lambda^2\theta(p)$. The point about these iterations is that they can be stopped (or even modified) after every expansion step. It is thus clear that one can group the counterterms that appear in the expansion together with the gates that get created when taking the average over the disorder. The cancellation among these two terms provides a small factor that makes such terms vanish in the diffusive scaling limit. Moreover, it is clear that one can implement rules for stopping the expansion independently of the subsequent disorder average. In particular, because the randomness is i.i.d., one can avoid moments beyond the power

$2d$ by stopping the expansion when a given site has appeared in the collision history d times. The terms to which no such repetition or renormalization cancellation applies are expanded up to order $n \sim \lambda^2 t \lambda^{-\delta} \sim \lambda^{-\kappa-\delta}$, where $\delta > 0$ depends on κ. The intuition behind this is that certain graphs with $n \sim \lambda^2 t \sim \lambda^{-\kappa}$ give the main contribution, and expanding up to an order that is $\lambda^{-\delta}$ higher leads again to small factors.

The imaginary part of ω gives effectively a regularization $O(\lambda^2)$ instead of $O(\eta)$ for the denominators, which changes the values of all diagrams significantly. In particular, the integral for one rung of the ladder becomes

$$\int \frac{\lambda^2 \, \mathrm{d}p}{(\alpha - \overline{\omega(p+r)} - i\eta)\ (\beta - \omega(p-r) + i\eta)} = 1 + C_0 \lambda^{1-O(\kappa)} \tag{37}$$

where C_0 is a constant. Thus with this renormalization, the ladders become of order 1, so that one can go beyond kinetic scaling. Indeed, in the language of Feynman graphs, the main result can be stated informally as

> After renormalization, the sum of the ladder graphs for the Wigner transform converges to a solution of the heat equation in the diffusive scaling limit.

The precise statements are Theorems 5.1, 5.2, and 5.3 in Ref.[7] They involve in particular proving that the terms which do not correspond to pure up–down pairings vanish in the limit, and dealing with a number of technical complications which arise from the fact that one has to do an expansion to a finite order.

3.6. *The key estimate for controlling combinatorics*

We have had to leave out almost all technical details to avoid overloading the presentation, but we should like to at least mention the heart of the proof here at the end, to clarify the main ideas about the Feynman graph expansion.

Focusing on up-down pairings, we have to deal with a combinatorial problem of bounding the sum over the $n!$ permutations $\sigma \in \mathcal{S}_n$. As mentioned, with an expansion to infinite order, one cannot get beyond the kinetic scaling because of this factor $n!$. The control of the remainders is done here by choosing an appropriate stopping n for the expansion and by "beating down the combinatorics by power counting". That is, we prove exponential suppression of the values of Feynman graphs in the number of crossings they have, that is, loosely speaking, in their complexity.

The precise notion capturing the complexity of a permutation $\sigma \in \mathcal{S}_n$ is its degree $d(\sigma)$, defined as the number of non–ladder and non–antiladder indices. Essentially, the ladder indices are those for which $\sigma(i+1) = \sigma(i)+1$, and the antiladder indices are those for which $\sigma(i+1) = \sigma(i) - 1$.

Theorem 3.1. *Let Γ_σ be the Feynman graph corresponding to σ. There is $\gamma > 0$ such that for all σ*

$$|Val(\Gamma_\sigma)| \leq C\lambda^{\gamma d(\sigma)}. \tag{38}$$

This theorem is proven using a special integration algorithm for bounding the values of large Feynman graphs.[7]

The number of permutations with degree D is

$$\mathcal{N}_{n,D} = |\{\sigma \in \mathcal{S}_n : d(\sigma) = D\}| \leq 2(2n)^D. \tag{39}$$

Expanding up to $n = O(\lambda^{-\kappa-\delta})$, $\delta > 0$, we have by (38), if $\gamma - \kappa - \delta > 0$,

$$\sum_{\substack{\sigma\in\mathcal{S}_n\\ d(\sigma)\geq D}} \lambda^{\gamma d(\sigma)} = \sum_{d=D}^{k} \lambda^{\gamma d}\mathcal{N}_{n,d} \leq 2\sum_{d=D}^{k}(2\lambda)^{d(\gamma-\kappa-\delta)} \leq O(\lambda^{D(\gamma-\kappa-\delta)}). \tag{40}$$

Thus the contribution from the sum of all terms with degree $D \geq 2$ is small if $\gamma - \kappa - \delta > 0$, hence the essential restriction for the value of κ is that of γ. As mentioned, one would hope to get close to $\gamma = 2$ in (38), but γ has to be chosen smaller for technical reasons.

4. Conclusion

We have shown that, for random Schrödinger operators with a weak static disorder the quantum mechanical time evolution can be controlled on large space and time scales where a diffusion equation governs the behavior. The Schrödinger evolution is time–reversible – yet irreversibility on large scales emerges. This apparent controversy is resolved by noting that along the scaling limit microscopic degrees of freedom have been effectively integrated out.

Although the expansion methods we use bear some resemblance to those of constructive quantum field theory, there are also a few noteworthy differences. First, because we analyze the time evolution at real time, the (near–)singularities of the propagators are located on hypersurfaces, and not at points, as would be the case in Euclidean field theories. The singularity structure is to some extent similar to that in real time Fermi surface problems, although there is no fixed Fermi surface here – the integrals over α and β "test" all possible level sets of the function $e(p)$, and this leads to a number of serious complications. Second, we are able to control the combinatorics of a straightforward Feynman graph expansion in momentum space, while the analysis in constructive field theory (to our knowledge, always) needs to be done by cluster expansions in position space to avoid the divergence of an infinite series of Feynman graphs. The reason for this is twofold: the unitarity bound allows us to do an expansion to a finite order, and our strong improvement (38) over standard power counting bounds allows us to push this order so high that we can reach the scale where diffusion sets in, while still retaining control of the remainders.

The genuine challenge is to show diffusion without taking scaling limits, i.e. for a fixed (small) disorder λ and for any time independent of λ. With expansion techniques, this would require to renormalize not only the self–energy to arbitrary order but also the four–point functions. Refining the self–energy renormalization poses no fundamental difficulty. The correct renormalization of all four–point functions in this problem, however, remains a widely open problem.

References

1. O. E. Lanford III,*On the derivation of the Boltzmann equation.* Astérisque **40**, 117-137 (1976)
2. H. Kesten, G. Papanicolaou: *A limit theorem for stochastic acceleration.* Comm. Math. Phys. **78** 19-63. (1980/81)
3. T. Komorowski, L. Ryzhik: *Diffusion in a weakly random Hamiltonian flow.* Commun. Math. Phys. **263** no.2. 277-323 (2006)
4. D. Dürr, S. Goldstein, J. Lebowitz: *Asymptotic motion of a classical particle in random potential in two dimensions: Landau model*, Commun. Math. Phys. **113** (1987) no 2. 209-230.
5. L. Bunimovich, Y. Sinai: *Statistical properties of Lorentz gas with periodic configuration of scatterers.* Commun. Math. Phys. **78** no. 4, 479–497 (1980/81),
6. D. Dürr, S. Goldstein, J. Lebowitz: *A mechanical model of Brownian motion.* Commun. Math. Phys. **78** (1980/81) no. 4, 507-530.
7. L. Erdős, M. Salmhofer and H.-T. Yau, *Quantum diffusion of the random Schrödinger evolution in the scaling limit.* Advances in Mathematics (2008) DOI 10.1007/s11511-008-0027-2
8. L. Erdős, M. Salmhofer and H.-T. Yau, *Quantum diffusion of the random Schrödinger evolution in the scaling limit II. The recollision diagrams.* Commun. Math. Phys. **271**, 1-53 (2007)
9. L. Erdős, M. Salmhofer and H.-T. Yau, *Quantum diffusion for the Anderson model in scaling limit.* Ann. Inst. H. Poincare **8**, 621-685 (2007)
10. P. A. Lee, T. V. Ramakrishnan, *Disordered electronic systems.* Rev. Mod. Phys. **57**, 287–337 (1985)
11. D. Vollhardt, P. Wölfle, *Diagrammatic, self-consistent treatment of the Anderson localization problem in $d \leq 2$ dimensions.* Phys. Rev. B **22**, 4666-4679 (1980)

12. A. Klein, *Absolutely continuous spectrum in the Anderson model on the Bethe lattice*, Math. Res. Lett. **1**, 399–407 (1994)
13. M. Aizenman, R. Sims, S. Warzel, *Absolutely continuous spectra of quantum tree graphs with weak disorder.* Commun. Math. Phys. **264** no. 2, 371-389 (2006)
14. R. Froese, D. Hasler, W. Spitzer, *Transfer matrices, hyperbolic geometry and absolutely continuous spectrum for some discrete Schrödinger operators on graphs.* J. Funct. Anal. **230** no 1, 184-221 (2006)
15. F. Germinet, A. Klein, J. Schenker, *Dynamical delocalization in random Landau Hamiltonians.* Ann. Math. **166** (2007) 215 – 344
16. J. Fröhlich and T. Spencer, *Absence of diffusion in the Anderson tight binding model for large disorder or low energy*, Commun. Math. Phys. **88**, 151–184 (1983)
17. M. Aizenman and S. Molchanov, *Localization at large disorder and at extreme energies: an elementary derivation*, Commun. Math. Phys. **157**, 245–278 (1993)
18. I. Rodnianski, W. Schlag, *Classical and quantum scattering for a class of long range random potentials.* Int. Math. Res. Not. **5** 243–300 (2003).
19. J. Bourgain, *Random lattice Schrödinger operators with decaying potential: some higher dimensional phenomena.* Lecture Notes in Mathematics, Vol. 1807, 70-99 (2003).
20. H. Spohn: *Derivation of the transport equation for electrons moving through random impurities.* J. Statist. Phys.**17** (1977), no. 6., 385-412.
21. L. Erdős and H.-T. Yau, *Linear Boltzmann equation as the weak coupling limit of the random Schrödinger equation*, Commun. Pure Appl. Math. **LIII**, 667-735, (2000).
22. T. Chen, *Localization Lengths and Boltzmann Limit for the Anderson Model at Small Disorders in Dimension 3.* J. Stat. Phys. **120** (2005), no. 1-2, 279-337.

RENORMALIZATION IN CHAOTIC AND PSEUDOCHAOTIC DYNAMICAL SYSTEMS

JOHN H. LOWENSTEIN*

Department of Physics, New York University
New York, NY 10003, USA
**E-mail: john.lowenstein@nyu.edu*

Renormalization played a central role in reviving classical dynamics as an exciting area of research during the second half of the twentieth century, and continues being an invaluable theoretical tool in the twenty-first century. Its defining characteristic is dynamical self-similarity, which allows one reliably to probe asymptotically small distances and asymptotically long orbits in phase space. We consider two applications of renormalization in nonlinear dynamics. The first is the historic discovery by Feigenbaum of the universal period-doubling route to chaos. The second is a more recent treatment of local and global scaling behavior of a periodically kicked harmonic oscillator. Some of these models exhibit the phenomenon of pseudo-chaos, where complex fractal structure in phase space is generated without the exponential divergence of nearby orbits which characterizes true chaos.

Keywords: Renormalization; Nonlinear dynamics; Dynamical systems.

1. Introduction

As is well known, renormalization tamed the infinities of quantum field theory and thereby led to its ascendency as the preeminent theory of elementary particles during the second half of the twentieth century. Perhaps less well known is the role of renormalization, during the same half-century, in breathing new life into the dormant field of classical dynamics, leading to exciting new understanding of nonlinear phenomena. The present paper is an attempt to explain and illustrate how renormalization has been employed to tame infinities in nonlinear dynamics, both historically and in current research.

The states of a classical system are typically points of a manifold (phase space) which evolve in time according to a dynamical mapping on the manifold. For example, a Hamiltonian system with n degrees of freedom has its states in a $2n$-dimensional phase space, with continuous time evolution specified by $2n$ differential equations. Other dynamical systems are inherently discrete. Even a continuous system, viewed stroboscopically, admits a discrete description of the time evolution, and so we shall assume below that a dynamical map exits.

In classical dynamics, renormalization is synonymous with *dynamical self-similarity*, referring to the following general scenario. Suppose a region of phase space $D(0)$ is mapped into itself by a dynamical map $\rho^{(0)}$. Let $D(1)$ be a proper sub-region of $D(0)$ which differs from it by a similarity transformation S (in the group generated by translations, rotations, inversions and scale transformations),

$$D(1) = S(D(0)).$$

Let $\rho^{(1)}$ be the first-return map on $D(1)$ induced by $D(0)$, i.e. for all $x \in D(1)$, there exists a minimal positive integer n, depending on x, such that

$$x' = \rho^{(0)\,n}(x) \in D(1).$$

We define

$$x' = \rho^{(1)}(x).$$

We have dynamical self-similarity (a renormalization process) if $\rho^{(1)}$ is just $\rho^{(0)}$ conjugated by S:

$$\rho^{(1)} = S \circ \rho(0) \circ S^{-1}.$$

Obviously, if this condition is satisfied, we have a countable sequence of nested phase-space domains $D(n) = S^n(D(0))$, as well as a sequence of

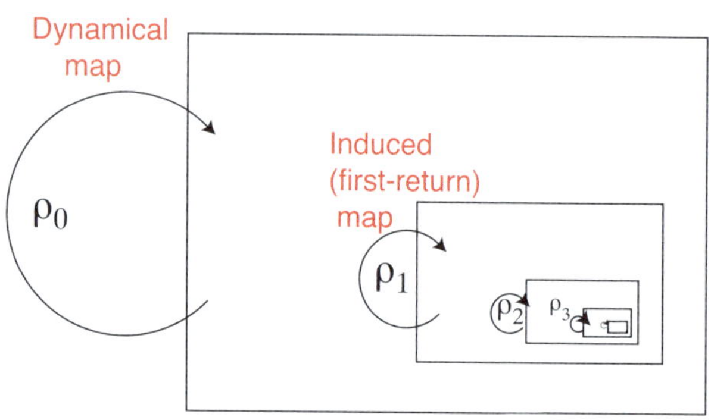

Fig. 1. Renormalization as dynamical self-similarity.

dynamically similar first-return maps $\rho^{(n)}$ implementing the nesting. The situation is depicted schematically in Fig. 1. The existence of a system with dynamical self-similarity may give rise to universal scaling properties in dynamical systems which differ from it by a small deformation. We will see how this operates in the example of the next section.

2. Period-doubling cascade

2.1. *Rayleigh-Bénard convection experiment*

A particularly instructive application of dynamical self-similarity in physics underlies the beautiful Rayleigh-Bénard convection experiment carried out by A. Libchaber, C. Laroche, and S. Fauve,[1] shown schematically in Fig. 2. In the experiment, liquid mercury is confined in a small rectangular box (28 mm x 7 mm x 7 mm), whose top and bottom are maintained at respective temperatures T_{low} and T_{high}. When the Rayleigh number R, proportional to the temperature difference $T_{\text{high}} - T_{\text{low}}$ is below a critical threshold R_c, heat is conducted from bottom to top in a uniform flow. As the Rayleigh number rises through R_c, cylindrical convective rolls make their appearance, and this can be verified by means of a small temperature probe inserted in the chamber. Increasing the Rayleigh number further produces transverse oscillations in the convective rolls, with these producing temperature fluctuations in a stationary probe. The latter were carefully recorded by the experimenters for increasing values of the Rayleigh number, with the results displayed in the figure. Immediately above threshold, the amplitude of the oscillations is steady, while for slightly higher values a period-doubling bifurcation occurs: the peaks alternate in height, corresponding to a repetition time twice as long. As R is increased further, a sequence of such bifurcations occur, with the difference between successive bifurcations shrinking at an exponential rate as one approaches a limiting value R_∞:

$$R_\infty - R_n \sim \delta^{-n}, \quad n = 4.4 \pm 0.1. \tag{1}$$

The Fourier transform of a periodic signal exhibits frequency peaks at integer multiples of the inverse of the period. Each period-doubling bifurcation in the Rayleigh-Bénard experiment is accompanied by the insertion of a new peak midway between nearest neighbors in the spectrum. The lowest ("subharmonic") frequency is one-half of the fundamental frequency. The observed frequency bifurcations in the experiment are displayed in Fig. 2 in the form of a power spectrum (absolute square of the Fourier amplitude).

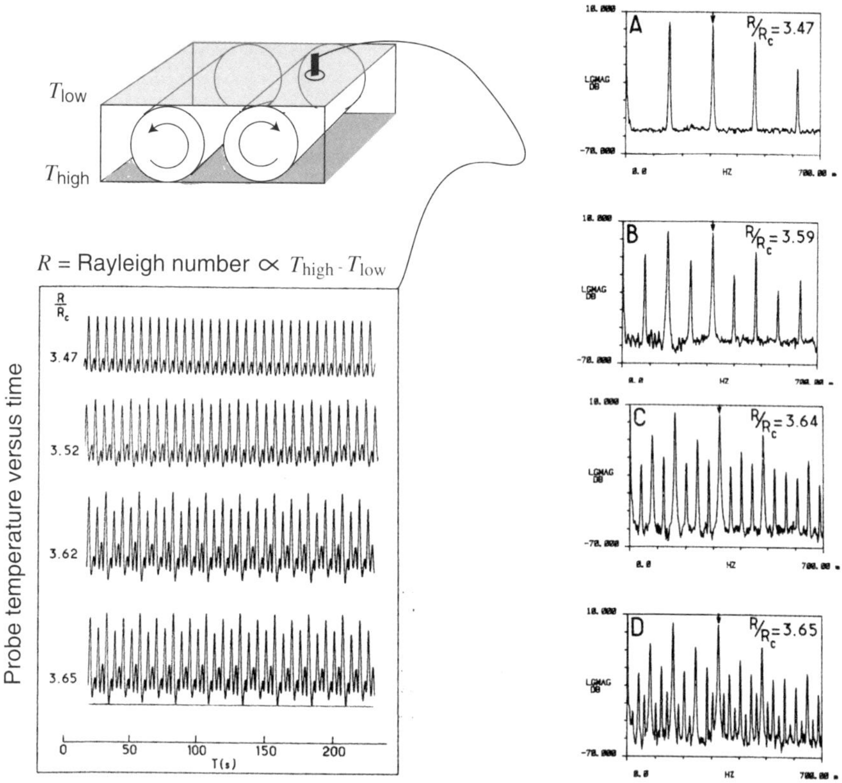

Fig. 2. The Rayleigh-Bénard convection experiment of Libchaber, Laroche, and Fauve.

As R approaches the period-doubling limit as in (1), the power spectrum shows pronounced scaling behavior of the peak amplitudes.

To understand the Rayleigh-Bénard scaling phenomena, we need not have an accurate mathematical model of all of the degrees of freedom of a specific fluid undergoing convection. It will be sufficient to relate the experimental system to a much simpler toy model– the one-dimensional logistic map! This miraculous linkage will be seen, of course, to be a consequence of renormalization.

2.2. *Logistic map*

A simple mathematical model of the period-doubling cascade is provided by the logistic map,

$$x \mapsto f(x) = \lambda x(1-x), \quad 0 \le x \le 1, \quad 0 \le \lambda \le 4. \tag{2}$$

The graph of $y = f(x)$ is shown in Fig. 3 (a), together with the diagonal $y = x$, for $\lambda = 2.9$. The two graphs intersect at the two points where $f(x) = x$, i.e. at fixed points of the map. The left fixed point, (0,0), is such that the slope $f'(0)$ is greater than one in magnitude, and hence it is a repellor: nearby points are mapped away from it. On the other hand the righthand fixed point, (x_*, x_*), has a slope $f'(x_*)$ which is negative and less than one in magnitude, and hence it is an attractor: nearby points approach (x_*, x_*), with the sequence ("time series") $x_1 - x_*, x_2 - x_*, x_3 - x_*, \ldots$ tending toward an alternating-sign geometric series.

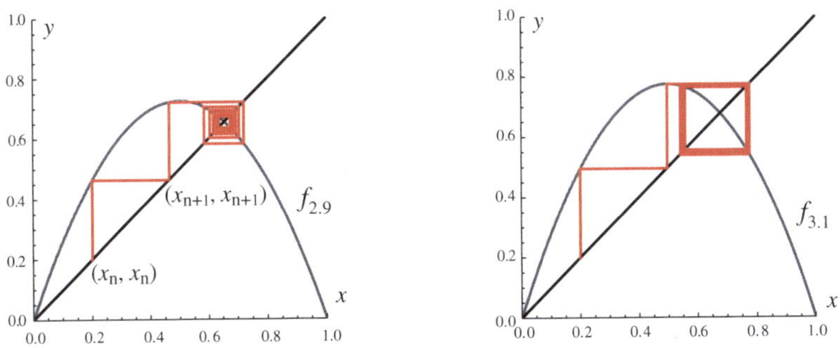

Fig. 3. Iteration of the logistic map.

A convenient way of following orbits of the logistic map is to concentrate on the sequence of points on the diagonal (x_n, x_n), generated by the simple rule: (i) draw the vertical line segment from (x_n, x_n) to $(x_n, f(x_n))$, then (ii) draw the horizontal segment from $(x_n, f(x_n))$ to $(f(x_n), f(x_n)) \stackrel{\text{def}}{=} (x_{n+1}, x_{n+1})$. (iii) iterate this prescription for $n = 0, 1, 2, \ldots$. In Fig. 3 we have applied the geometrical construction for $\lambda = 2.9$ and $\lambda = 3.1$. In the first case, the orbit is seen asymptotically to spiral into the righthand fixed point, whereas in te second case it tends toward a 2-cycle. If one increases continuously the parameter λ, one discovers a bifurcation point at $\lambda = 3$, precisely where the slope of $f(x)$ achieves the value -1. As the parameter approaches 3 from either side, the collapse to the attractor (fixed point or

2-cycle) takes longer and longer, the collapse time eventually diverging to infinity.

The parameter λ is analogous to R in the Rayleigh-Bénard experiment, and the bifurcation at $\lambda = 3$ is the first of an infinite sequence of period-doubling bifurcations. Instead of tracking the bifurcations, it is advantageous to locate, for each period 2^n, $n = 0, 1, 2, \ldots$, the particular value $\lambda = \lambda_n$ for which the orbit contains the point $x = \frac{1}{2}$. It is easy to see that for this "superstable" map, the convergence rate to the attractor is maximized.

In Figs. 4 and 5, are displayed the first 6 members of the period-doubling cascade, with respective periods $1, 2, 4, 8, 16, 32$. The plots of $x(t)$ versus the iteration number (discrete time) t may be compared with the time series for the temperature-probe readings in the convection experiment (Fig. 2).

High-precision numerical determinations yield arbitrarily precise values for the asymptotic parametric and geometric ratios as one approaches the

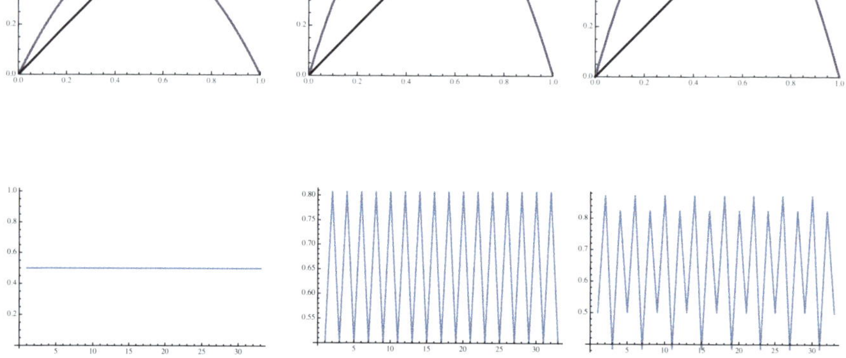

Fig. 4. Superstable periodic orbits of the logistic map.

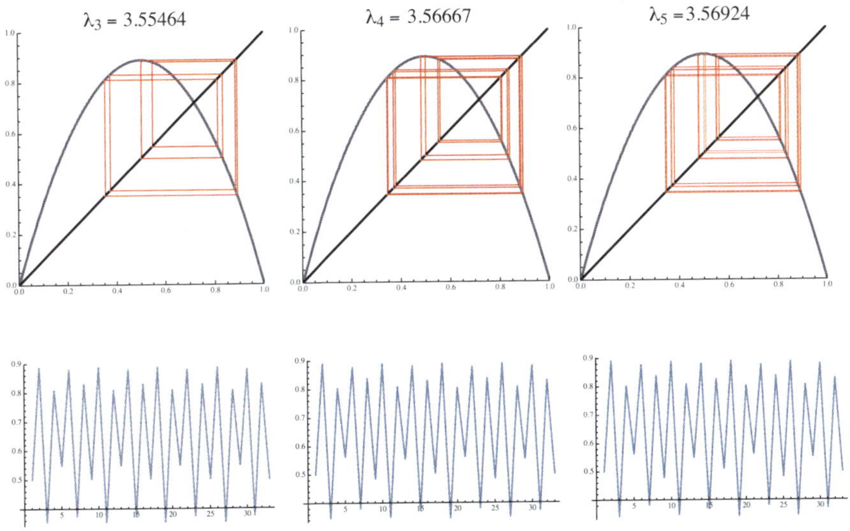

Fig. 5. Superstable periodic orbits of the logistic map, continued.

period-doubling limit λ_∞.

$$\lambda_\infty = 3.569945671..... \tag{3}$$

$$\delta = \lim_{n\to\infty} \frac{\lambda_n - \lambda_{n-1}}{\lambda_{n+1} - \lambda_n} = 4.669201609..., \tag{4}$$

$$\alpha = \lim_{n\to\infty} \frac{x_n(0) - x_n(2^{n-1}}{x_{n+1}(0) - x_{2^n}} = -2.502907875... \tag{5}$$

That the measured values of δ in the convection experiment agree, within experimental uncertainty, with those of the logistic map is no coincidence. What is going on here was discovered by Feigenbaum in his very first crude numerical explorations of period doubling. In addition to the function $\lambda x(1-x)$, he also studied a variety of functions (e.g. $\lambda \sin(x/\pi)$) which are smooth positive functions on the unit interval with a simple quadratic maximum, multiplied by a parameter λ. In each case he found a period doubling cascade with the same values of δ and α. These are then of universal significance within an infinite-dimensional universality class of functions. The implications are profound: the universality class appears to be so broad that it should contain some examples in the realm of natural phenomena, or at least in the experimental physics laboratory. In fact, the Libchaber et al. experiment was specifically designed to produce such an example.

Where does the universality come from? Fortunately, a thorough analysis of the logistic map allows one,[2,3] via a renormalization process, to penetrate to the heart of the matter. Here we shall merely sketch the argument, while pointing out that the main conclusions have indeed been backed up by mathematically rigorous proofs.[4]

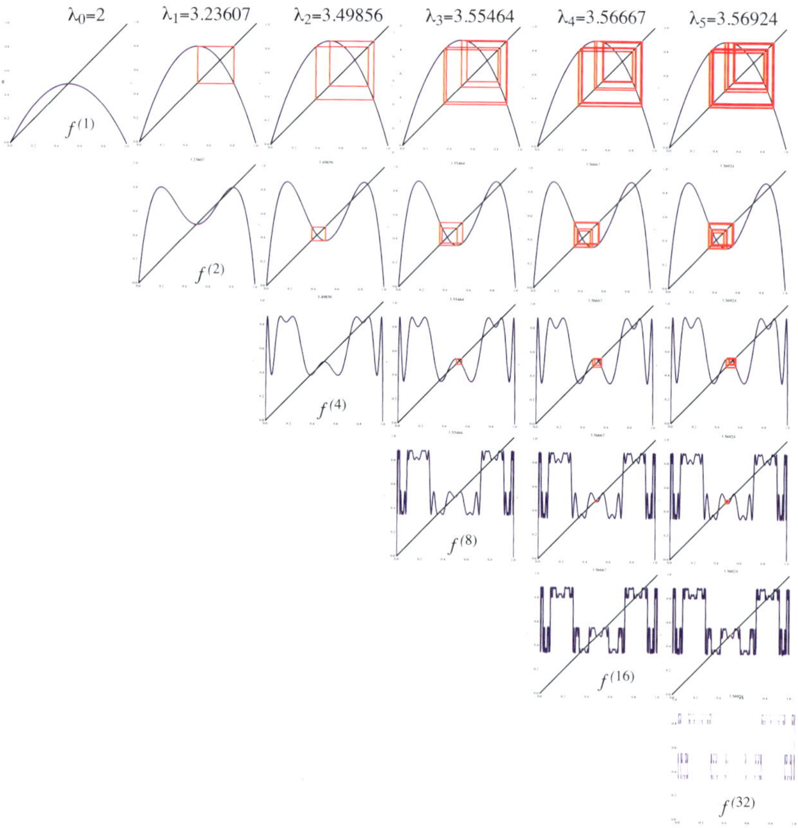

Fig. 6. Array of superstable orbits of the logistic map and its iterates. The period increases horizontally to the right, and the iteration number (a power of 2) increases vertically downward.

Consider the tableau of orbit plots of Fig. 6. The top row, row 0, consists of the first 6 superstable periodic orbits of the logistic map. The next row, row 1, consists of the first 5 superstable periodic orbits of f^2, the second iterate of f. In general, row n, contains the superstable periodic orbits of f^{2^n}. In Fig. 6 one readily spots a generalized form of renormalization (dynamical self-similarity). Consider, for example, the main diagonal of the array. Starting with row 0, column 0, namely the superstable fixed point, take one step down and to the right. In the plot of $f^2_{\lambda_1}$, one spots a central subinterval on which the induced dynamics is, to the eye, identical to that of $f^1_{\lambda_0}$, only inverted (relative to the point at $x = \frac{1}{2}$) and rescaled by a factor $\approx 1/\alpha$. The same relationship appears to hold for every step down the diagonal. What can be proved is that in the limit $n \to \infty$, the scaling becomes exact, and one approaches a limiting function g_0. A similar limiting process holds for each of the diagonals, as depicted in Fig. 7. This leads to an infinite sequence of functions g_n:

$$\lim_{n\to\infty} (-\alpha)^n f^{2^n}(\lambda_{n+k}, \frac{x}{(-\alpha)^n}) = \mu g_k\left(\frac{x}{\mu}\right),$$

with μ determined by the normalization of the infinite-n limiting function (see below):

$$\lim_{n\to\infty} g_n(x) = g(x), \quad g(0) = 1.$$

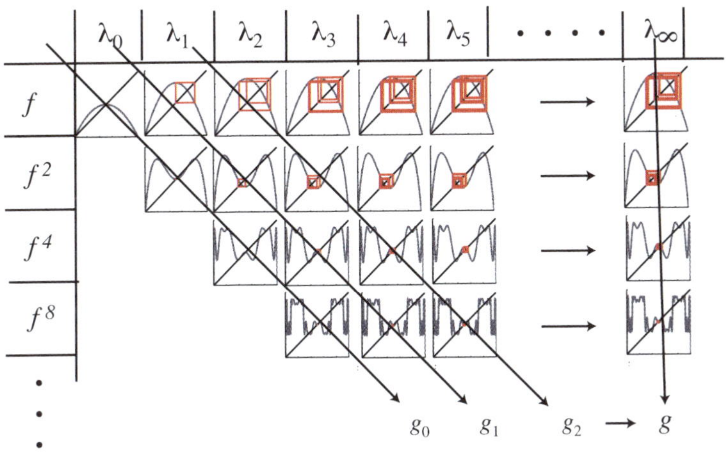

Fig. 7. Scaling sequences of superstable orbits of the logistic map and its iterates.

The dynamical self-similarity described above is slightly more general than the general definition of our introduction: here each step involves not only induction on a sub-region and a similarity transformation, but also a shift in the parameter defining the map. In our original sense, the truly renormalizable logistic map is the one with $\lambda = \lambda_\infty$, obtained by moving infinitely far to the right on our tableau (see Fig. 7). For this "column", each step downward consists of induction on a sub-interval (translated,inverted, and rescaled) which reproduces approximately the dynamics of the previous row. For the sub-interval, the first-return path consists of just two iterations of the previous map. In the limit of $n \to \infty$, the rescaling is by a factor $1/\alpha$, and there is a limiting function g (the Cvitanovíc-Feigenbaum function) which is invariant with respect to the renormalization step. It is the same function obtained by first calculating the g_n, then taking the limit $n \to \infty$.

It is important to recognize that although we have introduced the Cvitanovíc-Feigenbaum function g by means of a limiting process within the context of the logistic map, the function is a universal one which would have resulted from an analogous limiting process for any dynamical system (including experimental ones) within the broad universality class of sufficiently smooth maps with quadratic maxima. The situation is elegantly summarized in Fig. 8, which is a sketch showing g as a saddle-type fixed point in function space of the transformation T defined by

$$Th(x) \stackrel{\text{def}}{=} (-\alpha)h^2\left(\frac{x}{-\alpha}\right).$$

Thus the fixed-point condition takes the form

$$g(x) = Tg(x) = (-\alpha)g(g\left(\frac{x}{-\alpha}\right)),$$

with the normalization condition

$$g(1) = g(g(0)) = -\frac{1}{\alpha}.$$

We see that the geometric scale factor α is not an independent input, and we can redefine T to be independent of α. Thus,

$$g(x) = Tg(x) \stackrel{\text{def}}{=} g(1)^{-1}g^2(g(1)x).$$

We note that the stable manifold, consisting of all those functions which converge to g under iterated application of T, is an infinite dimensional surface, while the corresponding unstable manifold is a one-dimensional curve along which T acts, asymptotically near g, as multiplication by Feigenbaum's $\delta = 4.669...$. The universal functions g_k are points on the unstable manifold satisfying

$$g_k(x) = Tg_{k+1}(x).$$

In the diagram, the logistic map is represented by a curve f_λ which intersects the stable manifold of g transversely at f_{λ_∞}. Iterated application of T gives a sequence of maps converging to g (corresponding to descending the rightmost column of Fig. 7).

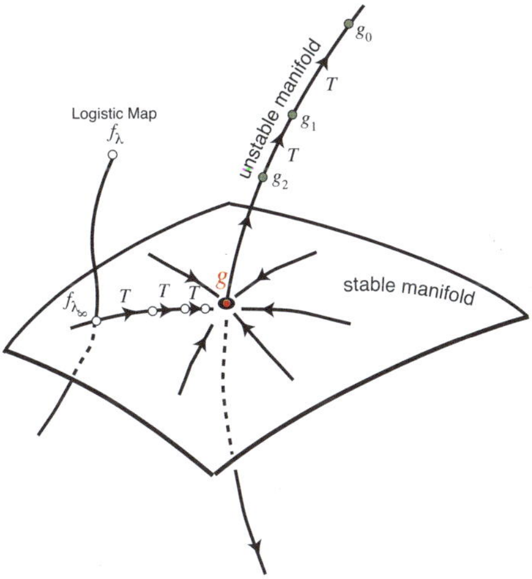

Fig. 8. Sketch of the function-space fixed point g, together with its stable and unstable manifolds.

3. Chaotic and pseudochaotic kicked oscillators

We now consider a class of simple dynamical systems exhibiting complicated long-time behavior resembling a diffusion process. In these systems renormalization clearly plays a leading role in understanding the long-time asymptotics, as well as the small-scale geometric structure in phase space (the two are in fact intimately related). The dynamical self-similarity will once again be expressed in terms of functional relations analogous to the Cvitanovíc-Feigenbaum equation, but they will be much more complicated and their lifting to function space much more obscure (and still unexplored).

Consider a one-dimensional harmonic oscillator (position x, momentum y, unit mass) which is kicked impulsively 4 times per natural period, with the amplitude of the kick a periodic function of x. Such a system may be described by a Hamiltonian

$$H(x,y) = \frac{1}{2}(x^2+y^2) + F(x)\sum_n \delta(t-\pi n/2),$$

where the derivative F' satisfies $F'(x) = F'(x + 2\pi)$. The equations of motion are then

$$\dot{x} = \frac{\partial H}{\partial y} = y, \tag{6}$$

$$\dot{y} = -\frac{\partial H}{\partial x} = -x - F'(x) \sum_n \delta(t - \pi n/2). \tag{7}$$

Thus, the motion of a point mass is an alternation of free oscillation for a quarter-period and a momentum shift

$$y \to y + \Delta y, \quad \Delta y = -F'(x).$$

This is illustrated in Fig. 9 for the case $\Delta y = K \sin x$ studied extensively by Zaslavsky and collaborators starting in the 1980's,[5] in part motivated

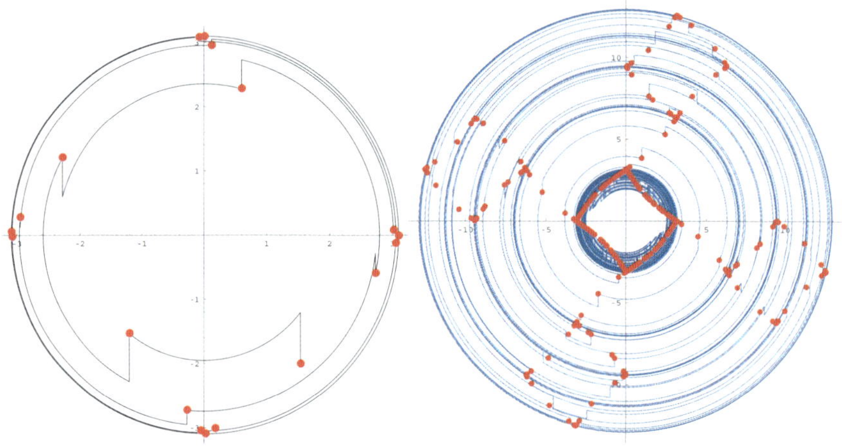

Fig. 9. Orbit of the kicked-oscillator with kick function $0.8 \sin x$. The stroboscopic orbit (Poincaré section) is marked with solid dots.

by applications in plasma physics and the physics of fluids. In the figure, the left-hand frame shows a partial orbit consisting of 17 kick-periods, each represented by a momentum shift (vertical displacement $0.8 \sin x$) followed by a quarter-circle traversed clockwise. To view the orbit stroboscopically, we have placed a solid dot at the start of each kick period. Successive dots are related by the discrete Poincaré map

$$W \begin{pmatrix} x \\ y \end{pmatrix} = \begin{pmatrix} 0 & 1 \\ -1 & 0 \end{pmatrix} \begin{pmatrix} x \\ y + K \sin x \end{pmatrix}.$$

The longer orbit to the right in Fig. 9 shows the advantage of the stroboscopic view: a curvilinear quadrilateral is being traced out near the origin, with rotated, translated replicas further out in the plane. Increasing the iteration number to 40,000, but recording only every fourth point, we see in Fig. 10 that the particular orbit we have chosen continues to move outward,

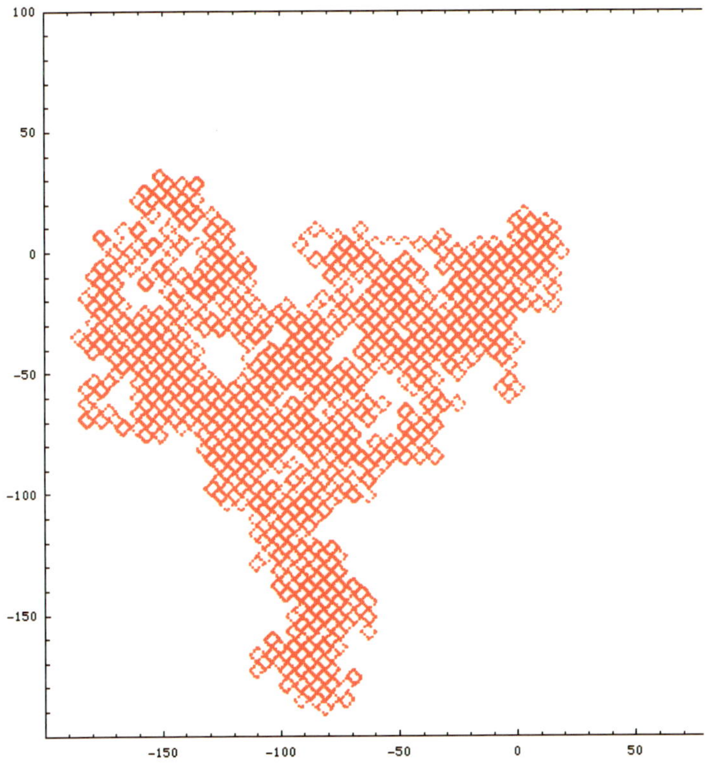

Fig. 10. Ten thousand iterations of W^4.

filling out a web-like region with apparent 4-fold crystalline symmetry in the plane. This orbit has been chosen from a neighborhood of chaotic initial conditions: i.e. nearby orbits diverge from one another at an exponential rate. In addition, the orbits tend to wander around the plane executing what appears to be a random walk. The region occupied by these orbits, shown in Fig. 11, is known as a stochastic web.

To what extent does the average long-time behavor of the chaotic orbits

Fig. 11. Stochastic web. The highlighted rectangle is the fundamental cell.

in the stochastic web mimic a genuine random walk? The answer turns out to depend on the value of the parameter K.[6] For typical values, the orbits of W^4 in the stochastic web proceed to infinity with mean-square distance from the initial point satisfying, for time t tending to infinity,

$$< x^2 + y^2 >\sim D\, t$$

where D plays the role of a diffusion constant. However, for special values of K, the righthand side is better described by a super-diffusive power law, i.e.

$$< x^2 + y^2 >\sim D'\, t^{\mu},$$

with $\mu > 1$.

By means of a high-precision numerical experiment, Zaslavsky and Niyazov[6] were able to determine the dynamical mechanism for the anomalous diffusive behavior. It is essentially one of dynamical self-similarity (renormalization). Because of the 4-fold crystalline symmetry of the model, it pays to consider, in addition to the web orbit wandering to infinity, the folded orbit obtained by relacing each point (x, y) by $(x \bmod 2\pi, y \bmod 2\pi)$ in the fundamental cell of the "crystal". The super-diffusive parameter values are those for which there are "accelerator modes" – points which are periodic with respect to the local map, but which, which lie on global orbits which tend to infinity with linear velocity – and these accelerator modes are surrounded by a dynamically self-similar system of island chains (islands around islands). The orbits consist of inter-cell flights interspersed with long sojourns in the island systems. The net effect is a super-diffusive power law, where the power μ can be related to the spatial and temporal scale factors of the island-around-island hierarchy.

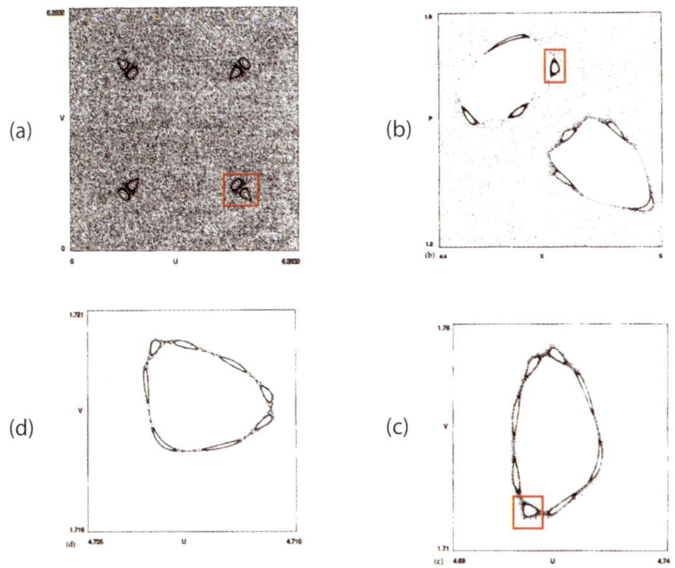

Fig. 12. Island-around-island orbit of the stochastic web map.

One such island-around-island system is shown in Fig. 12. Frame (a) shows the entire fundamental cell for the special value $K = 6.349972$. The islands (magnified in (b)) surround accelerator modes, and are immersed in a chaotic sea occupying most of the phase space. Note that the density of plotted points in (b) is much higher at the island boundaries, indicating trapping there. In (c), one of the islands of (b) has been magnified, revealing that it, too, is actually a chain of islands, with elevated density of orbit points at their boundaries. Further magnifications, such as in (d), show a hierarchy of island chains, with both spatial and temporal scaling properties. Of course, the special value of K was not known at the outset, but rather was tuned recursively to produce trapping around the islands of each level.

The behavior one sees at the boundary of islands is not truly chaotic, since nearby orbits, while capable of branching profusely, cannot separate at an exponential rate. Such behavior has been termed pseudo-chaotic by Zaslavsky. In recent years, my collaborators and I have been interested in exploring dynamical models in which all of the orbits outside of periodic or quasi-periodic islands are pseudo-chaotic. Within the class of kicked oscillators this can be achieved by replacing the sinusoidal kick function in the stochastic web map by a piecewise linear one, i.e. a sawtooth function, as in Fig. 13.

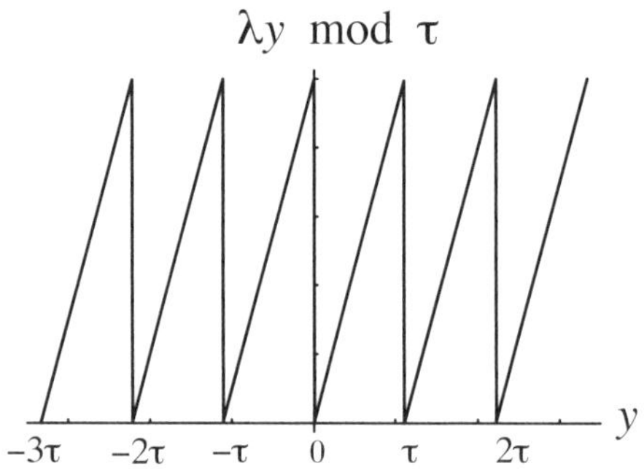

Fig. 13. Sawtooth kick function.

In the notation of,[7] we write the global (Poincaré map on the infinite plane) and local (folded map on the fundamental cell) maps, respectively denoted W and K, as

$$W\begin{pmatrix} x \\ y \end{pmatrix} = \begin{pmatrix} 0 & 1 \\ -1 & 0 \end{pmatrix}\begin{pmatrix} x+\lambda y \\ y \end{pmatrix} = \begin{pmatrix} y \\ -x-\lambda y \end{pmatrix}$$

and

$$\begin{pmatrix} x \\ y \end{pmatrix} = \begin{pmatrix} y \\ -x-\lambda y \bmod \tau \end{pmatrix},$$

where τ is the period of the sawtooth function. For the model with $\lambda = \tau^{-1} = -\sqrt{2}$, which we shall use as our example, the actions of the global and local maps are depicted in Figs. 14 and 15, respectively.

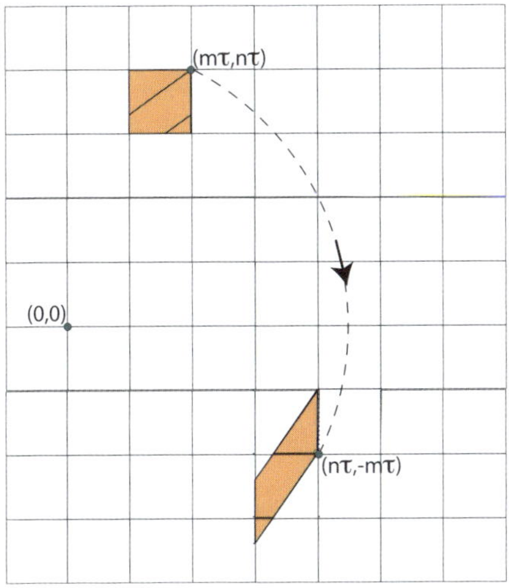

Fig. 14. Action of the global map W on a single cell.

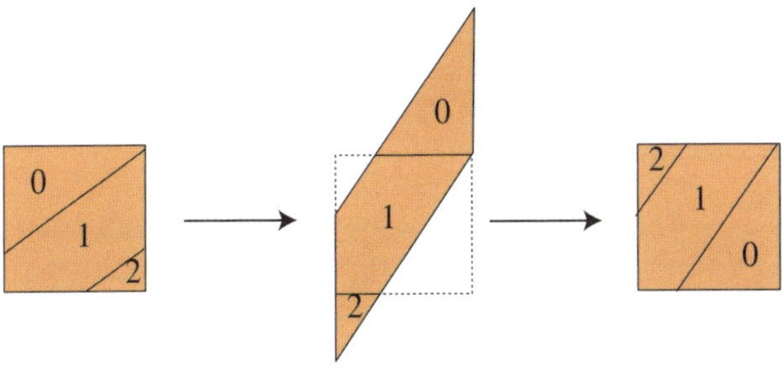

Fig. 15. Action of the local map K on the fundamental cell.

Any point in the plane can be labeled by a point u in the square $[0, \tau)$ (if $\tau > 0$) or $(\tau, 0]$ (if $\tau < 0$), and an integer pair $z = (m, n)$ labeling the relevant cell in the infinite lattice. The action of W then decomposes as

$$W(u + z\tau) = K(u) + L_u(z), \tag{8}$$

where $L_u(z)$ is the lattice isometry

$$L_u(z) = (F \cdot z)\tau + d(u), \qquad F = \begin{pmatrix} 0 & 1 \\ -1 & 0 \end{pmatrix}.$$

with $d(u)_x = 0$ and $d(u)_y$ taking one of the values $1, 0, -1$ depending on which of the three subdomains of the fundamental square (labeled 0,1,2 in the figure) u belongs to.

The fact that the lattice orbits generated by iteration of L_u depend at each step on the local coordinate vector u, suggests that we first understand the local orbits and their scaling properties, and then apply that knowledge to understand global scaling and asymptotic power laws. The

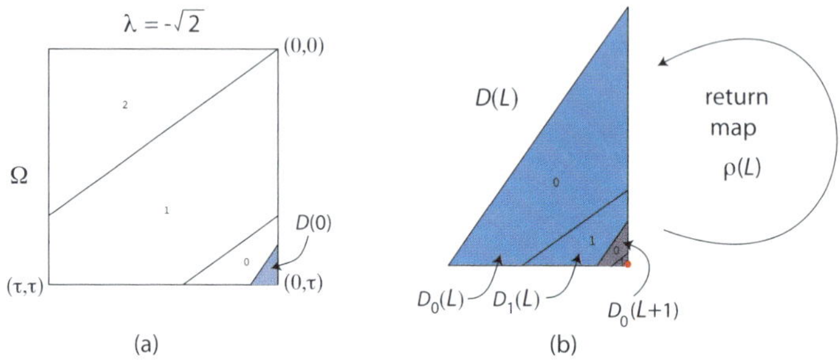

Fig. 16. Dynamical self-similarity of the nested sequence of triangles $D(L)$.

local renormalization structure is shown in Fig. 16. In the lower righthand corner of the fundamental cell Ω, there is a small right triangle $D(0)$ with the following properties:

(1) $D(0)$ is the disjoint union (up to boundary line segments) of two subdomains, $D_0(0)$ and $D_1(0)$, each of which returns intact to $D(0)$ after a finite number of iterations of K. This is $\rho(0)$, the first-return map on $D(0)$ induced by K.

(2) $D(0)$ is the first member of an infinite nested sequence of similar right triangles $D(L)$, such that $D(L+1)$ is identical to $D(L)$ rescaled by a factor ω (the same for all L), and the first-return map $\rho(L+1)$ induced by $\rho(L)$ is just a rescaled version of $\rho(L)$.

(3) For each L, the $\rho(L)$ first-return orbits of the subdomains $D_0(L+1)$ and $D_1(L+1)$ cover $D(L)$, up to boundary line segments and periodic domains. See Fig. 17.

(4) For each L, the return orbits of $D_0(L)$ and $D_1(L)$ induced by K cover Ω up to boundary line segments and periodic domains. See Fig. 18.

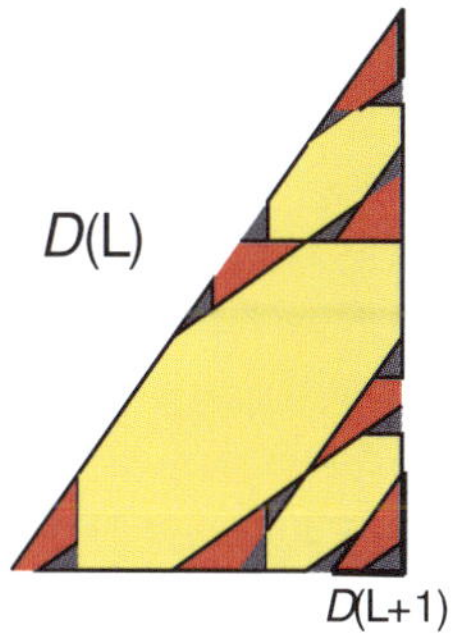

Fig. 17. Tiling of $D(L)$ by the first-return orbits of $D_0(L+1)$ and $D_1(L+1)$

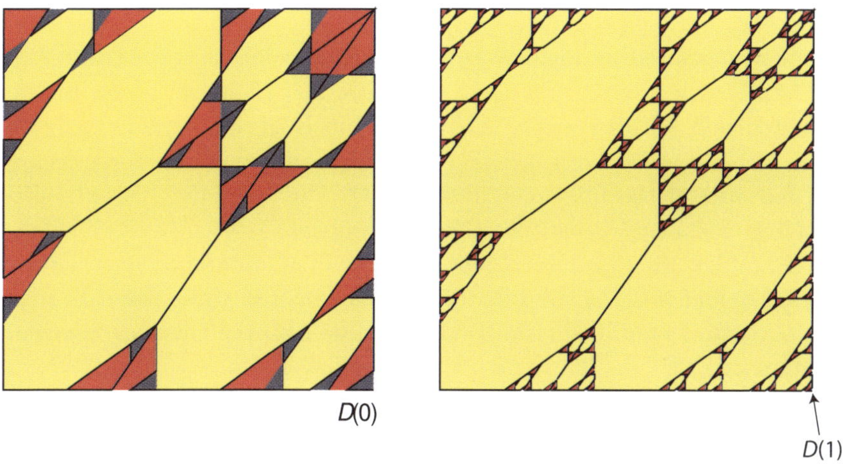

Fig. 18. Tiling of the fundamental cell by K orbits of sub-domains of $D(0)$ (left) and $D(1)$ (right).

The listed properties can be formulated in a language analogous to that of the Cvitanovíc-Feigenbaum equation for the period-doubling universal map. The geometrical self-similarity of the scaling sequence of triangles is expressed by

$$D_j(L+1) = S_{\omega_K} D_j(L), \quad j = 0, 1,$$

where the similarity transformation S is a rescaling by ω_K relative to the corner point $(0, \tau)$. The dynamical self-similarity expresses the fact that the first-return map $\rho(L+1)$, restricted to sub-domain $D_j(L+1)$, is a rescaled version of $\rho(L)$, restricted to $D_j(L)$. If the ν_j-step return orbit of $D_j(L+1)$ is

$$\left(D_{p(j,0)}(L), D_{p(j,1)}(L), \ldots, D_{p(j,\nu_j-1)}(L)\right),$$

then $\rho(L+1)$ restricted to $D_j(L+1)$ is

$$\rho_j(L+1) = \rho_{p(j,\nu_j-1)}(L) \circ \cdots \circ \rho_{p(j,1)}(L) \circ \rho_{p(j,0)}(L) = S_{\omega_K} \circ \rho_j \circ S_{\omega_K^{-1}}.$$

Although this equation is analogous the Cvitanovíc-Feigenbaum equation, its significance is, thus far, not nearly as far-reaching, for the simple reason that we have not solved the highly nontrivial problem of embedding the problem in a suitable function space so that our return map may be viewed as the fixed point of a renormalization group flow. If that were possible, we could identify a universality class of models, all with the same asymptotic scaling properties.

As is suggested by Fig. 18 (b), the fraction of phase space covered by the first-return orbits to the level L triangle $D(L)$ induced by K, shrinks monotonically with increasing L. More and more of the polygonal periodic domains are revealed with each increment of L, and eventually these occupy the full area of the fundamental square Ω. The aperiodic, discontinuity-avoiding orbits then constitute a zero-measure set Σ.

If $T_j(L)$ is the first-return time (number of K iterations) for the orbit of $D_j(L)$, and if A_{ij} is the L-independent number of times the $\rho(L)$ first-return orbit of $D_j(L+1)$ visits $D_i(L)$, then we have the following recursion relation,

$$T_j(L+1) = \sum_{k=0}^{\nu_j-1} T_{p(j,k)}(L+1) = \sum_i T_i(L) A_{ij}.$$

Asymptotically for large L, the first-return time is dominated by ω_T^L, where the temporal scale factor ω_T is the largest eigenvalue of the transpose of the matrrix A. It is not difficult to calculate fractal (Hausdorff or box-counting) dimension of Σ,[7] namely

$$\dim(\Sigma) = \frac{\log \omega_T}{|\log \omega_K|}.$$

It would be nice if the asymptotic long-time behavior of the kicked-oscillator orbits initiated in Σ were simply related to its fractal dimension. The actual situation turns out to be more complicated, thanks to the presence of the global 4-fold lattice rotation. Roughly stated, simple power-law asymptotics will emerge only if there is a suitable synchronization of the local and global "clocks".

To understand the phenomenon of global scaling leading to asymptotic power laws, let us recall the local-global decomposition (8) of the kicked-oscilator map W, which we conveniently rewrite in terms of a complex coordinate for the lattice point (m, n), $\zeta = n + mi$.

$$W([u, \zeta]) = [K(u), i\zeta + \delta(u)], \qquad \delta(u) \in \{1, 0, -1\}$$

Note that the $\pi/2$ rotation is now represented by multiplication by $i = \sqrt{-1}$.

We next introduce $\rho_W(L)$, which is the local first-return map $\rho(L)$ lifted to the entire plane. For $u \in D_j(L+1)$, $\zeta \in \mathbf{Z} + \mathbf{Z}i$, we have the recursion relation

$$\rho_W(L+1)([u, \zeta]) = [\rho(L+1)(u), i^{T_j(L+1)}\zeta + \delta_j(L+1)],$$

where

$$\delta_j(L+1) = i^{\kappa(L,j,0)} d_{p(j,0)}(L) + i^{\kappa(L,j,1)} \delta_{p(j,1)}(L) + \cdots + \delta_{p(j,\nu_j-1)}(L),$$

with

$$\kappa(L, j, t) = \sum_{k=t+1}^{\nu_j - 1} T_{p(j,k)} \bmod 4.$$

Collecting terms, we get

$$\delta_j(L+1) = \sum_{k=0} M_{jk}(L)\delta_k(L),$$

where $M(L)$ is a matrix with Gaussian integer entries. If, for sufficiently large L, the matrix $M(L)$ becomes independent of L, then the asymptotic global scaling is governed by its largest-magnitude relevant eigenvalue, ω_W, and we can expect a power law with exponent

$$\mu = \frac{\log \omega_W}{\log \omega_T}.$$

The local and global scaling properties of kicked-oscillator models have been investigated for those cases where the parameter λ is a quadratic algebraic number of magnitude less than 2,[7] as well as one much more complicated case of a cubic irrational parameter.[8] A summary of the quadratic

Table 1. Summary of local and global scale factors for the quadratic kicked-oscillator models.

λ	ω_K	ω_T	ω_W	μ	behaviour
$\sqrt{2}$	$3-2\sqrt{2}$	9	1	0	bounded
$-\sqrt{2}$	$3-2\sqrt{2}$	9	1	0	bounded
$(1+\sqrt{5})/2$	$(3-\sqrt{5})/2$	4	1	0	bounded
$(1-\sqrt{5})/2$	$(3-\sqrt{5})/2$	4	1	0	bounded
$(-1+\sqrt{5})/2$	$(3-\sqrt{5})/2$	4	4	1	ballistic
$(-1-\sqrt{5})$	$(3-\sqrt{5})/2$	4	4	1	ballistic
$\sqrt{3}$	$7-4\sqrt{3}$	25	4	.430677	sub-diffusive
$-\sqrt{3}$ (A)	$2-\sqrt{3}$	4	2	.5	diffusive
$-\sqrt{3}$ (B)	$2-\sqrt{3}$	5	2	.4306770	sub-diffusive
$-\sqrt{2}$ (A)	$3-2\sqrt{2}$	9	1	0	logarithmic
$-\sqrt{2}$ (B)	$3-2\sqrt{2}$	9	5	.732487	super-diffusive

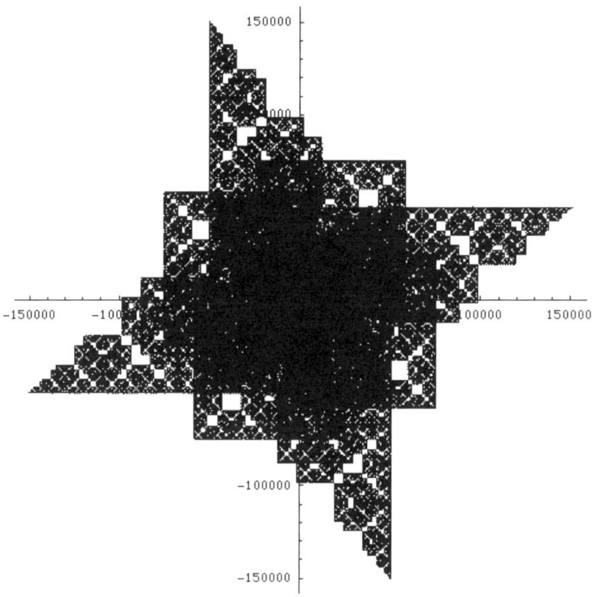

Fig. 19. Aperiodic orbit for $\lambda = -\sqrt{3}$ (1.4×10^6 iterations of $\rho(0)$).

results is shown in Table 1. We conclude this section with an interesting example. In Fig. 19, we show part of a subdiffusive aperiodic orbit for $\lambda = -\sqrt{3}$. This orbit slowly makes its way outward in the plane, sporadically returning to the neighborhood of the origin. Note the asymptotic

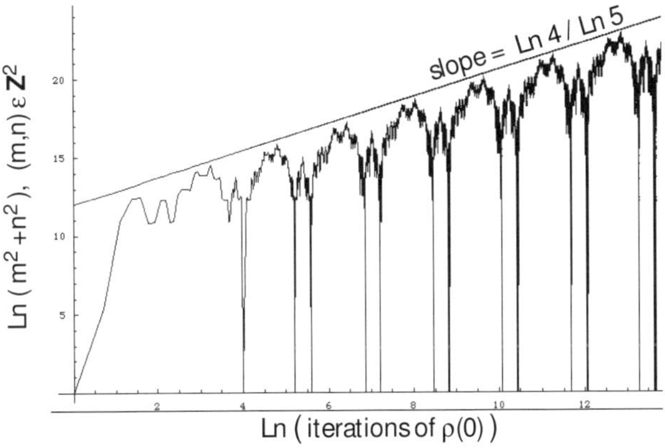

Fig. 20. Log-log plot of lattice distance squared versus the iteration number.

fractal structure of the log-log plot of distance vs. time in Fig. 20.

References

1. A. Libchaber, C. Laroche, and S. Fauve, *J. Physique Lett.* **43**, L-211 (1982).
2. M. J. Feigenbaum, *J. Stat. Phys.* **19**, 25 (1978).
3. M. J. Feigenbaum, *J. Stat. Phys.* **21**, 669 (1979).
4. O. E. Lanford, *Bull. Am. Math. Soc.* **6**, 427 (1982).
5. G. M. Zaslavsky, *Physics of Chaos in Hamiltonian Systems*, Imperial College Press, London, 1998.
6. G. M. Zaslavsky and B. A. Niyazov, *Physics Reports* **283** 73 (1997).
7. J. H. Lowenstein, G. Poggiaspalla, and F. Vivaldi, *Dynamical Systems* **20**, 413 (2005).
8. J. H. Lowenstein, Pseudo-chaotic orbits of kicked oscillators, Singapore lectures, preprint (2007).

AUTHOR INDEX